Preserving the Glory Days

T0289485

BOOKS BY SHAWN HALL

Romancing Nevada's Past: Ghost Towns and Historic Sites of Eureka, Lander, and White Pine Counties

Old Heart of Nevada: Ghost Towns and Mining Camps of Elko County

Preserving the Glory Days: Ghost Towns and Mining Camps of Nye County, Nevada

Preserving the Glory Days

Ghost Towns and Mining Camps of Nye County, Nevada

SHAWN HALL

 University of Nevada Press   Reno/Las Vegas

University of Nevada Press, Reno, Nevada 89557 USA

Copyright © 1981 by Shawn Hall

New material copyright © 1999 by University of Nevada Press

All photographs in the Shawn Hall collection copyright © 1999
by Shawn Hall, unless otherwise noted

Manufactured in the United States of America

Library of Congress Cataloging-in-Publication Data

Hall, Shawn, 1960–

[Guide to the ghost towns and mining camps of Nye County,
Nevada]

Preserving the glory days : ghost towns and mining camps of
Nye County, Nevada / by Shawn Hall.

p. cm.

"Edition expanded and revised from the 1981 publication"—
Publisher's info.

Includes bibliographical references and index.

ISBN 978-0-87417-317-8 (pbk. : alk. paper)

1. Nye County (Nev.)—History, Local.   2. Nye County (Nev.)—
Guidebooks.   3. Ghost towns—Nevada—Nye County—History.
4. Mining camps—Nevada—Nye County—History.   5. Ghost
towns—Nevada—Nye County—Guidebooks.   6. Mining camps—
Nevada—Nye County—Guidebooks.   I. Title.

F847.N9H34   1998

98-24318

917.93'34—dc21

CIP

The paper used in this book meets the requirements of American
National Standard for Information Sciences—Permanence of
Paper for Printed Library Materials, ANSI Z39.48-1984. Binding
materials were selected for strength and durability.

This book has been reproduced as a digital reprint.

To WILLIAM "BILL" METSCHER,

whose influence has strengthened my resolve to record Nevada's history.

I will always be grateful to have him as my friend.

Contents

Preface   ix

Acknowledgments   xi

Introduction   xiii

Northwestern Nye County   1

Northeastern Nye County   83

North Central Nye County   125

Central Nye County   185

Southern Nye County   233

References   275

Index   291

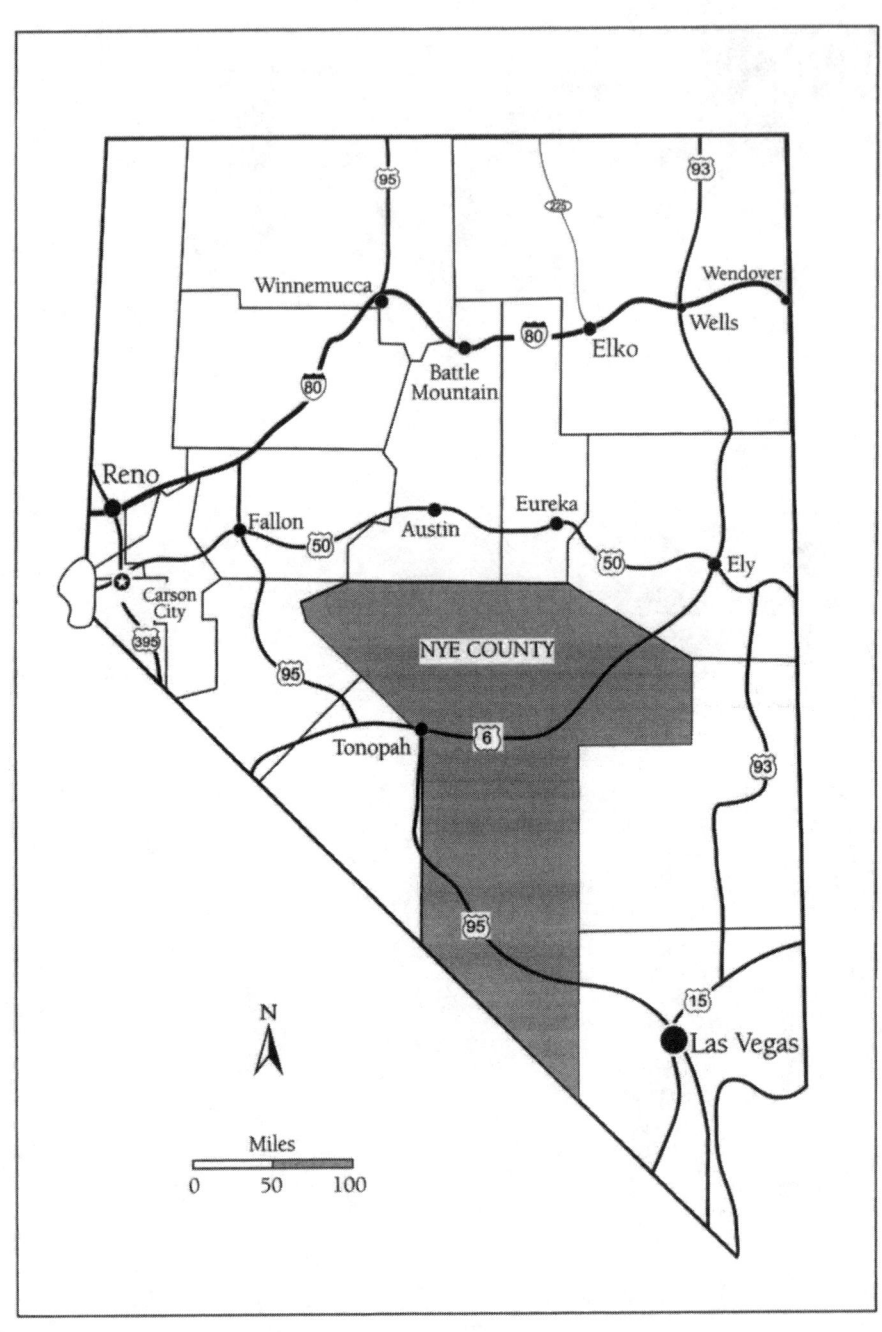

Winnemucca

95

93

225

Wendover

Wells

80

Elko

Battle
Mountain

Reno

Fallon

Austin

Eureka

50

50

Ely

Carson
City

NYE COUNTY

395

95

6

Tonopah

93

95

15

Las Vegas

N

Miles

0   50   100

Preface

Since I wrote *A Guide to the Ghost Towns and Mining Camps of Nye County, Nevada* in 1981, two additional volumes of my ongoing county-by-county Nevada series have been published. They are *Romancing Nevada's Past: Ghost Towns and Historical Sites of Eureka, Lander, and White Pine Counties* and *Old Heart of Nevada: Ghost Towns and Mining Camps of Elko County.* During the past fifteen years I have amassed a great deal of information on Nye County that did not appear in the first Nye County book, which has been out of print for some time. This volume, while it includes most of the material contained in the first, has been expanded by many pages. Not only is there additional information on the places covered in the first book, but it also contains more than twenty sites that were not mentioned in that volume.

As one might expect, over the past fifteen years readers have had the occasion to point out some errors in the first Nye County book, and these have been corrected in this volume. During the past two years I visited the ghost towns and mining camps of Nye County once again in order to provide updated information about current conditions of the towns I describe. This endeavor proved disappointing at times, because much of what I had seen in the late 1970s was gone. I expected some deterioration caused by the forces of nature to occur over the years, but most of the destruction I witnessed was the result of vandalism and not of the elements. A striking example of this type of senseless destruction is the razing of the Cosmopolitan at Belmont. It is unfortunate that an irresponsible few can ruin the opportunity for the rest of us to visit these sites for enjoyment.

This book is organized as a historical atlas and is designed both for the armchair reader and for the active ghost-town visitor. Though the distances to each town may not jibe precisely with the numbers on your vehicle's odometer, every site is marked on the map at the front of the book. Following the primary name of each site are other names by which it was known during its existence.

Be cautious while traveling in Nye County. Most of the area is very hot and dry, and water is not always readily available. Carry extra water and gasoline. Rattlesnakes abound in the county, so be careful when you are poking around. Also potentially dangerous are the many unmarked open mine shafts. While most shafts have been marked since I first visited Nye County, there are still a large number that remain unmarked and hazardous.

Enjoy the wonderful ghost towns and mining camps of Nye County. They are rich in history. But be warned: once you begin visiting ghost towns, you are in danger of becoming infected with ghost-town fever, an incurable but enjoyable affliction.

Acknowledgments

Many people were instrumental in the preparation of the original Nye County book as well as of this updated and expanded version. I would first like to thank my daughter Heather Ashley Hall, the light of my life. My parents, Albert and Lorraine Hall, have been a constant source of support and inspiration. They have encouraged me to pursue my dreams, even though it has meant living far away from their home in Massachusetts.

Neither version of this book would have been possible without the friendship and assistance of William Metscher, who has been supportive of my endeavors since I met him in 1978 while I was preparing the first volume on Nye County. In addition to lending me his historical expertise, he is also responsible for obtaining copies of many of the photographs that appear in the book. I will forever be in his debt. His brothers, Allen and Philip, also played a large part in seeing this project through to its completion. The three brothers are more dedicated than anyone I know to preserving the history of central Nevada. Due to their efforts, the Central Nevada Museum, which is one of the best local history facilities in the West, is now a reality. Visitors to the museum will find ample evidence of the Metscher brothers' love of Tonopah and the surrounding area.

I would also like to thank a number of other people who helped by providing me with information and friendship while I was engaged in the preliminary research for both this and the first Nye County volume. They are Mrs. LaRue Carter, Austin Wardle, Leila Fuson, Nick and Linda Bradshaw, Bob Work, Leland Hendrix, A. D. Hopkins, Jr., Jim Marsh, Charles Palmquist, Sharon Pauley, Bill Phillips, Helen Uhalde, Brooke Whiting, Dorothy Wilson, Julie Zimmerman, Bud Easton, Ray and Cheryl Kretschmer, Joe and Sue

Fallini, Helen Fallini, Sonja DeHart, Eslie Cann, Judith Riptoe, Heidi LaPoint, Mati Stephens, Belmont Bob, Charlie and Lois Chapin, Lyle Weir, Wayne Phillips, Wally Cuchine, Howard and Terry Hickson, Bruce and Gina Franchini, Patrick Connolly, Christina Ulm, Robert Nylen, Gloria Harjes, Nancy Marshall, Doug and Cindy Southerland, and the staff of the Central Nevada Museum. It saddens me that some of these friends have passed on since I worked with them from 1978 to 1981, but I will never forget them.

Special thanks go to Wayne Hage, who allowed this greenhorn Easterner to become a part of the Pine Creek work force during my first visit to Nye County in 1978. That and the subsequent five summers spent at the ranch were the most memorable times of my life. Working on the ranch not only fulfilled a childhood fantasy but also gave me a more accurate idea of what the Old West was really like.

I wish to thank all the people mentioned above as well as anyone else I might have missed. I look forward to seeing everyone once more when I travel through Nye County again in the years to come.

Introduction

Nye County was organized in 1864 and was named after the popular Nevada governor J. W. Nye. The first major discoveries in the county were made in 1863 in the Ione area of the Union District. Soon other towns sprang up nearby. They were Washington, Grantsville, Ophir City, and Belmont. A small courthouse was constructed in Ione at a cost of $800, but it wasn't until 1874 that a substantial courthouse was built at the new county seat, Belmont.

Nye County's boundaries constantly shifted during its early years. After a number of disputes, surveyors finally settled the boundaries. In 1870 Nye County had a reported population of 1,100. By 1875 this number had grown to 2,100. Growth halted in the early 1880s, and the mines in the county remained relatively dormant until the turn of the century, when Jim Butler made his rich discoveries at Tonopah Springs. This awakened keen interest in the old mines and sent prospectors scurrying over the land searching for new deposits. New towns sprang up at Rhyolite, Manhattan, Round Mountain, New Reveille, and Johnnie. The teens were the richest years for the county, with Tonopah leading the boom. Activity slowly declined. It picked up again during World War II.

After the war, most of the mines closed, although major operations continue today at Round Mountain, Manhattan, and Rhyolite. These modern-day huge open-pit operations are the backbone of Nye County's economy. The new technology of microscopic gold has created a new boom, which has taken the place of the molybdenum and barite boom of the 1970s and early 1980s. With extended deposits, it appears that the gold operations will be viable for many years.

The county's fascinating cross section of history gives it its varied beauty.

There are no active railroads in the county today, but the remains of its four lines—the Bullfrog-Goldfield, the Tonopah and Goldfield, the Las Vegas and Tonopah, and the Tonopah and Tidewater—still make for interesting exploration. In addition to the ghost towns and the present towns, the county includes many old stage stations. The stage roads are now faint but are still traversable and are an excellent way to go back in time to the era of the stagecoach. Nye County also has many areas of abundant natural beauty. The splendor of the Jefferson Mountains is awe-inspiring. Sites such as the Northumberland Cave offer up one of the rarest natural wonders a traveler will ever lay eyes upon. The county's many gorgeous canyons are also majestic. Ride up a canyon road for only a few miles and witness a transformation from desert and sagebrush into a land of aspens, clear and cool water, birds, and rocky splendor. The scenery, the people, and the many wonderful ghost towns make Nye County one of the most beautiful areas in the state.

Northwestern Nye County

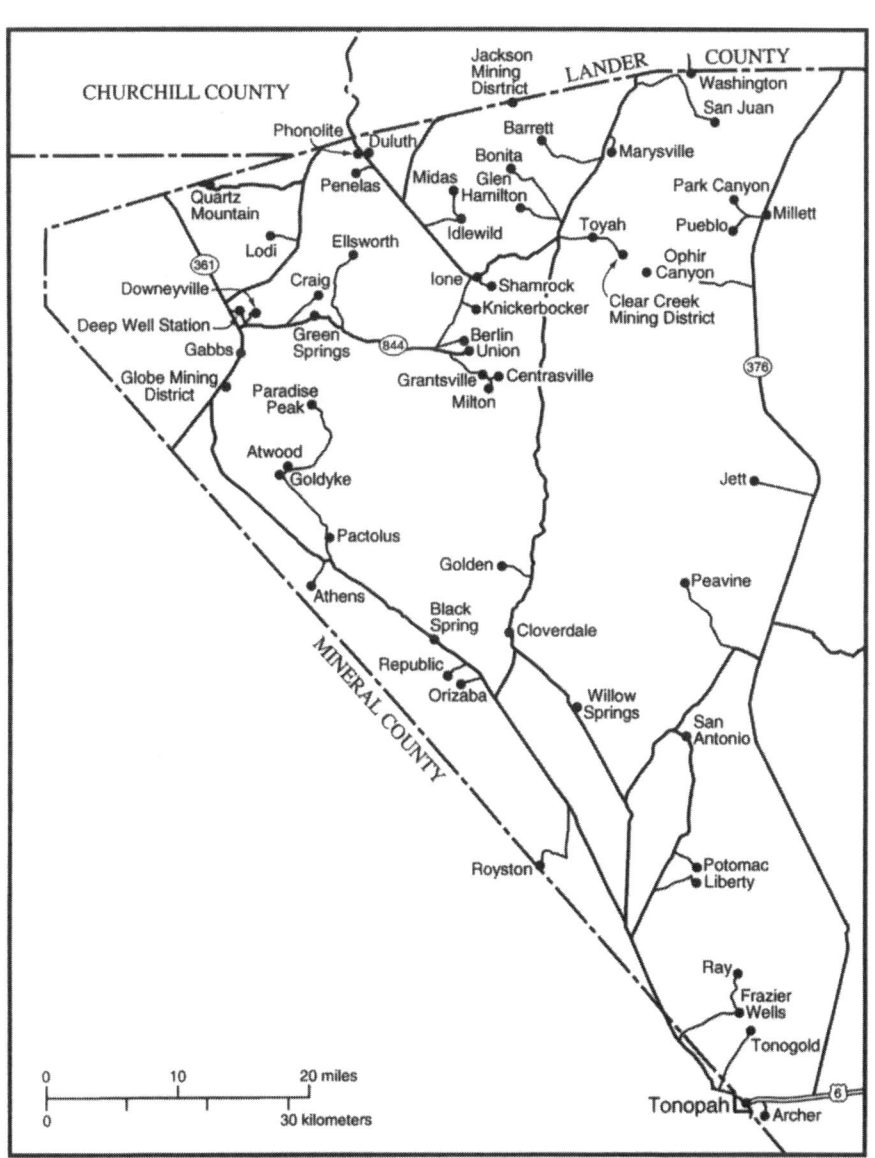

CHURCHILL COUNTY

Jackson Mining Disrtict

LANDER COUNTY

Washington
San Juan

Phonolite
Duluth
Penelas
Quartz Mountain
Lodi
Ellsworth
Craig
Downeyville
Deep Well Station
Gabbs
Green Springs
Globe Mining District
Paradise Peak
Atwood
Goldyke
Pactolus
Athens

Barrett
Bonita
Midas
Glen Hamilton
Idlewild
Ione
Shamrock
Knickerbocker
Berlin
Union
Grantsville
Milton
Centrasville

Marysville

Park Canyon
Pueblo
Toyah
Ophir Canyon
Clear Creek Mining District

Millett

376

Jett

Golden
Peavine
Black Spring
Cloverdale
Republic
Orizaba
Willow Springs
San Antonio

MINERAL COUNTY

Royston

Potomac
Liberty

Ray
Frazier Wells
Tonogold

Tonopah
Archer

6

361

844

0   10   20 miles

0   30 kilometers

Archer

DIRECTIONS: Archer is located 2 miles southeast of Tonopah.

Archer was a short-lived mining camp that sprang up during the summer of 1904 after some gold float was discovered there. No accompanying ledges were located, however, and the little tent camp had disappeared by the fall. No frame buildings ever went up at Archer, and nothing remains to show it once existed.

Athens

(Juniper Springs) (Warrior Mine)

DIRECTIONS: From Black Spring, head north on old Nevada 89 for 10.7 miles. Exit left and follow the road for ½ mile. At the fork, bear right and follow the road for 2 miles to Athens.

The Athens camp grew up around the Warrior mine and marks the center of the mining area. In 1910, even before the discovery of the Warrior mine, Athens experienced a short boom after John Martinez, J.R. Stott, and James Herald discovered rich ore at the site in 1909. By July 1910 there were seventy-five men at Athens, and the Athens Mining District formed. Miners erected thirty tents, Lester Bell platted a townsite, and lots sold for $75 for a corner and $50 for an inside. Businesses included a lodging house, a store, and a saloon.

A rival camp, Juniper Springs, sprang up nearby at the site of N.E. Dyer's discoveries. Dyer, who also owned the townsite, gave away free lots. The first businesses at Juniper Springs were the Juniper Lodging House and Guy Eckley's store. Eventually the camps merged, retaining the name of Athens. By the end of the month there were fifteen frame structures in Athens, and a tri-weekly stage, which W.E. McHaffy and S.C. Steele operated, ran from Athens to Mina. Stott and Martinez discovered a new deposit that assayed at $1,000 a ton, and John McGee optioned six of their claims for a total of $100,000. Unfortunately, however, the quality of the ore was disappointing, and the boom went bust by the end of the year. Under the headline "In Memoriam," the Manhattan *Post* mourned Athens's passing: "Athens, fair Athens, thy glory is gone. The burro has departed and the prospectors flown. And the gloom of the night shows no light in the street, for the boys have admitted you are dead to your feet."

A Civil War veteran who had been grubstaked by A.B. Millett, a store owner in Manhattan, originally discovered the Warrior mine, which was the district's principal producer. The discovery revived Athens and led to the construction of the first permanent buildings in the town. The Warrior Gold

Mining Company purchased the mine. Harry McNamara, who also owned a productive mine in the district, built a small stamp mill in 1913 and employed eighteen workers. During the first few months of production, the mines produced gold bullion worth $20,000. In 1921 the Aladdin Divide Mining Company bought the Warrior mine. The company then began extensive prospecting of the area and uncovered some new ore deposits, but by July it had already sold out to the Olympic Mines Company, which began shipping ore to its mill at Omco. Another company—the Lucky Boy Divide Mining Company, of which McNamara was president—bought the land adjacent to Athens and began active mining in 1922.

Mining activity continued at Athens for a few more years. Finally the ore grew too scarce, and operations shut down. In December 1931 J.J. McNeil, who brought in new equipment and reopened the mine, bought all the Warrior Company's property. In 1932 he sold out to Warrior Consolidated Gold, whose president was H.A. Houser. The company was unsuccessful, and L.B. Spencer and L.J. Smith bought the property in November 1933 at a sheriff's auction. Leasers continued to work the area until 1939, sending the ore they mined to the Dayton Consolidated mill in Silver City. No real production or additional activity took place at Athens after the leasers gave up in 1939.

Ruins still remain at Athens. The Warrior mine and a few buildings struggle to stand, and the site is covered with rubble. The foundation of a small amalgamation mill, built in 1913, is just east of the Warrior mine.

Atwood
(Okey Davis) (Fairplay) (Butler) (Gilt Edge)
(Edgewood)

DIRECTIONS: *From Gabbs, head south on Nevada 361 for 4 miles. Exit left and follow Sheep Canyon Road for 9 miles to Atwood.*

Atwood formed in 1901 after George Duncan, William Regan, E.A. McNaughton, and Okey Davis discovered gold in the surrounding hills. The camp became the center of the Fairplay Mining District, which miners organized in June 1903. In April 1904 the Gold Crown Mining Company was incorporated and began extensive work on the Atwood and Lone Star mines. Two years later, in September 1905, a townsite was platted, and a Tonopah realty company became the chief promoter of the 250-person town. Edgewood, a rival townsite, was platted nearby on December 26, 1905, but it soon disappeared.

In January 1906 the Griggs Atwood Mining Company purchased the town of Atwood, which became a company town. A post office opened at the camp on February 6, 1906. Henry Urquhart was postmaster, but Minnie Hovey soon replaced him. Hovey, who also owned the Atwood Hotel—a ten-room building with a lobby, a kitchen, and a dining room—moved the post office to the hotel. Other businesses in Atwood included the Bell Meat Market and the Urquhart Store. A dance hall went up, and residents organized a baseball team, playing against a team from nearby Goldyke on Sundays. Atwood's principal mines were the Butler, owned by Butler; the Atwood, owned by Griggs; and the Golden Crown, owned by Everett. Miners shipped ore to Sodaville. By the end of 1906 the town had a population of 200. J. Holman Buck, who had previously owned the *Nevada Copper News* at Acme, published a newspaper, the Fairplay *Prospector,* for a short while in 1907. The cost of a subscription was $4 a year, but the paper didn't last that long.

Atwood's mainstay was the Butler mine, which had a 280-foot shaft and 300 feet of drift work. When this mine closed in 1908 most of the interest in the Atwood area had faded. The post office closed on January 31, 1908, and the town was completely abandoned soon afterwards.

A revival of the town took place beginning in 1914, when Okey Davis discovered rich ore just south of the old townsite. Miners erected a small camp of eight buildings at the mine, and some people moved back to Atwood. The Butler mine, renamed the Nevada Chief, reopened, and a separate camp formed, named Butler after M. L. Butler, the discoverer of the Nevada Chief mine. In January the Nevada Chief Mining Company, owned by O. O. Emmons, Bob Williams, and Charles Steele; the Nevada Chief Extension Mining Company, which owned the Crackerjack, the Gold Coin, the Chief Fraction, the North Pole, and the South Pole mines; the Contact Mining Company, owned by J. R. Richardson; and the Excelsior Twilight Mining Company were working in the district. Jim Skelton organized the Butler townsite in February, and the first frame house was built. Soon a lodging house owned by A. Fredina, a cookhouse, and a lumber and feed yard were in business. By March the camp had a population of seventy-five, but the boom went bust during the summer. Residents renamed the dying camp Gilt Edge, but it was abandoned by fall. Atwood suffered the same fate.

The Oatman United Gold Mining Company, of which Gordon Booth was president, operated the Okey Davis mine, sinking a 300-foot shaft on the property. In 1927 the company also took over the Nevada Chief mine. But the mines ceased to produce, and by the early 1930s all activity had stopped. As late as 1950 most of the buildings at the mines were still standing. Walter Pfefferkorn, the last resident of the district, lived at the Okey Davis camp until he finally left in 1959. Today only piles of wooden rubble mark the Butler and Okey Davis sites. At Atwood, only one foundation and some broken glass are left.

Barrett

DIRECTIONS: From Ione, head north on old Nevada 21 for 17 miles. Exit left and follow the road for 1½ miles to a fork. Take the left road and follow it for 1¼ miles to Barrett.

Barrett, a small stage stop, was active during the late 1870s and early 1880s. The station was named after James Barrett of Georgia, who ran the station. He came to Nye County in 1869, a few years before he set up the Barrett stage stop. The station was built at the high point of Barrett Canyon, and drivers used it as a place to change horses after the hard climb to the summit. A solid wood structure and a few small cabins, along with a complement of horse stables, stood at the station. A small restaurant operated there, and lodging was available. A post office opened at Barrett on May 24, 1882, with Barrett as postmaster.

The need for stages running north from Ione diminished as that town faded. By 1884 none were passing through Barrett. The post office closed on January 23, 1885, and the stop was abandoned soon afterwards. Some limited mining took place in Barrett Canyon beginning in the late teens, when James Ward and his family began working a mine and built a small mill there. The mine and mill's modest proceeds supported the family, which stayed in Barrett until 1939, when Ward's wife died and he left.

A great deal of searching is required to locate the scattered wood scraps marking the remote site, which is situated at the end of the road. The road tends to be very rough and is impassable during wet weather.

Berlin

DIRECTIONS: From Ione, head south on Nevada 844 for 3 miles. Exit left and follow signs to Berlin for about 2 miles.

According to most historical accounts, State Senator T.J. Bell discovered the Berlin mine in 1895 and later sold the mine property to John Phelps Stokes of New York in 1898. However, in March 1887 the Belmont *Courier* reported that the Cincinnati Mining Company had opened the Berlin mine and was treating the ore at its Knickerbocker mill. It appears that Bell didn't actually discover the mine but rather relocated it, since the Cincinnati Company abandoned the mine when it folded in 1889.

The town of Berlin was founded in 1897. In 1898 the Nevada Company, which Stokes organized, bought two mills in the Ione area and transferred the equipment to Berlin for installation in a new thirty-stamp mill that was being built at the time. A post office, with John Thompson as postmaster,

opened on July 10, 1900. Berlin kept growing, and by 1905 the town had a population of close to 300.

Berlin was the site of a large store that had been moved there from Ione in 1900. The store also served as a boardinghouse and was one of the most impressive structures in the town. In addition to the saloons located in nearby Union Canyon, there were three others closer to Berlin. The Johnson Saloon moved to Ione in 1910, but owner Bob Johnson continued to run another establishment in Berlin until 1919. The Puccinelli Saloon was situated halfway between Berlin and Union Canyon. During the early years of Berlin's existence, this saloon was the most popular one in the area, because it was closest to the mine and mill. Billy Bell, who had earlier bought the Puccinelli Saloon, built the Bell Saloon—the last one to go up in Berlin. Bell inherited $3,000 and decided to use the money to build a new saloon in an excellent location just below town. Business was brisk at the Bell until Berlin was abandoned. In 1910 Bob Dixon moved the Union Canyon Town Hall to Berlin to serve as a boardinghouse, and a stage line also ran.

In 1907 everything went dead when a miners' strike closed down the mine and mill. The company refused to cave in, telling the miners it couldn't afford higher wages. It eventually folded. A few people remained, and their persistence was rewarded in late 1909, when leasers named Parman and Feenaman reopened the mine and mill. The modest revival, which only lasted a year, brought some life back to Berlin. Soon after Parman and Feenaman's operation shut down, Alfred Smith took a lease and constructed a fifty-ton cyanide plant just below the mill. The plant continued to operate on a small scale from 1910 until 1914, and the recovery rate of the cyanide processing was just $2.50 a ton.

After the Smith plant shut down, no more activity took place in the Berlin area for a number of years. The post office closed on December 18, 1918. Postmasters who served included Oren Counsil, Thomas Jones, Thomas Stevens, Charles Houstand, Daniel Johnson, and Lawrence Tanner.

Berlin was in its final death throes. The Goldfield Blue Bell Mining Company's purchase of claims in the area in the 1920s shone a small ray of hope over the district. The company reequipped the Berlin mine with a steam hoist and an air compressor, but the operations were of an exploratory nature and were never full scale. The company kept mine superintendent Daniel Johnson in Berlin until 1947, when the mill was dismantled. There has been no mining activity in the area since then, and the total value of the district's production was recorded at $850,000.

There are extensive remains in Berlin. Thirteen buildings are left at the site, which has been incorporated into the Berlin-Ichthyosaur Nevada State Park. The state park system has placed the old town of Berlin in a state of "arrested decay." Among the buildings still remaining are the huge mill (which the state recently restored to working condition), the old assay office (which now serves as a ranger's station), and a number of miners' cabins. All three saloons have disappeared, but an explorer can easily locate the building sites by looking for piles of broken glass. The ruins of five buildings, including the Puccinelli Saloon, are half a mile south of town. There is a small cemetery southeast of Berlin, but most of its graves are now empty, having been moved to different towns after the Berlin collapse. The ghosts of Berlin rest quietly. Only occasional visitors disturb their peace.

Berlin is a definite must for any ghost-town lover. Many hours can be spent wandering around the site. Signs throughout, erected by the state park system, label and describe the buildings still standing and others that are completely gone.

Black Spring

DIRECTIONS: *From Cloverdale, head south for 1 mile. Exit right and follow the road for 4 miles. Exit right onto old Nevada 89 and follow it for 3.1 miles to Black Spring.*

Black Spring was the scene of fairly heavy mining, but not of silver or gold. Diatomaceous earth was the "mineral" mined. The Nature Products Company, headed by J.M. Fenwick, purchased well over 200 acres of the mineral-bearing earth, which contained a high concentration of silica as well as small amounts of iron oxide, alumina, and soda. The minerals obtained from the soil were used as a base for Super Dent Tooth Powder and Super Dental Cream. Nothing of consequence remains as a reminder of the activity that took place in the area, but Black Spring still gives forth a cool stream of water.

Bonita

DIRECTIONS: *From Ione, take old Nevada 21 north for 11 miles to Bonita Canyon.*

An obscure stage station that was active during the early 1900s, Bonita was a stop on the Ione-Austin stage line. On August 2, 1907, a post office opened in Bonita, mainly as a result of a flurry of mining activity in the canyon. William Snyder was postmaster. Active companies in the area were the Reese River Gold Ledge Mining Company and the Reese River Gold Mining Company.

Miners made the first discoveries at Bonita in July 1906, and a small camp quickly formed. By September the population stood at 150, but no lasting ore deposits were located, and almost everyone was gone by the time winter hit. The post office closed for good on March 19, 1908. Soon the stage line stopped running, and Bonita faded into the past. Some cinnabar was mined there in 1912, but little other activity has taken place since. Nothing whatsoever remains at the station site, and the road leading to it is very rough and eventually vanishes. The station was situated in the canyon, but it was not possible to determine its exact location.

Centrasville

DIRECTIONS: *Centrasville is located 3½ miles east of Grantsville.*

Centrasville was a small camp that formed in 1878 after John Centras made a discovery there and began working the Monterey and George Martin mines. Centras had earlier been active in the mines at Tybo. By 1880 about fifteen men were employed in Centras, and residents erected a few buildings, including a boardinghouse, in the town. No businesses ever opened in Centrasville, however, because of the camp's proximity to Grantsville.

At this time Harry Gates started working the Hoodlum mine, but interest in the area faded as ore values dropped. By 1886 only Centras and a couple of other miners were left. In August Centras received some local notoriety when it was reported that he had spotted the "Reese River Serpent," which had long been a legend with local Indians. Centras said that the serpent was forty feet in length.

By 1890 even Centras had given up on his mines. In 1896 P.M. Bowler reopened the Monterey mine and employed sixteen men, but by 1898 he too had given up. Activity at Centrasville had come to an end. Now only mine dumps and some wood debris mark the site.

Clear Creek Mining District
(Crane Canyon)

DIRECTIONS: *From Ione, continue northeast for 7 miles. Exit right and head southeast for 2 miles. This is Clear Creek. To get to Crane Canyon, exit left and follow the road for 2 miles.*

The Clear Creek Mining District formed in 1907 after discoveries were made in both Clear Creek and Crane Canyons in May 1906. The district was located opposite Ophir on the other side of the Toiyabe Mountains. Over the next couple of years many mines began operating there. They were the Snow King, owned by Duncker and Olholf; the Great Western, owned by G.N. Gooding; the Big Chief, owned by Charles Feutsch; the Discovery, owned by Nathan Shapiro; the Right Tip, owned by Bert Virden and John Schmaling; the Silver Star, owned by Charles Wenderhof; and the Mayflower, owned by Wenderhof. The Toiyabe Mining Company and the King Midas Mining Company were also active in the district. However, low ore values and small deposits kept any serious development from occurring. By the mid-teens very little had been produced, and all the mines had been abandoned. Only small mine dumps are left in the two canyons.

Cloverdale

DIRECTIONS: *From Gabbs, head south on Nevada 361 for 5.2 miles. Exit left on old Nevada 89 and follow it for 30.2 miles. Exit left and follow the road for 3¾ miles. At four corners, take the left branch, and follow the road for 1 mile to Cloverdale.*

Cloverdale was a stage station on the Belmont wagon road in the 1860s. The Cloverdale Ranch was established during this time, and it became quite prosperous. In 1872 more stages were routed through Cloverdale. The ranch owners were known to be very hospitable, greeting many a weary traveler as they would a friend. In 1880 the northward flow of traffic through Cloverdale was heavy enough to give rise to thoughts of building a narrow-gauge railroad from Ledlie on the Nevada Central Railroad near Austin. Plans for the Nevada Southern Railroad vanished after it was determined that the project was not financially feasible. In 1880 Adam Farrington built a new large stone station at Cloverdale to accommodate the increased number of travelers. In the 1880s and 1890s T. J. Bell made his home at Cloverdale and ran numerous stage lines throughout Nye County and beyond. Bell was also active in mining and politics, serving many years as an assemblyman and, in 1893, as Speaker of the House.

A post office, with Mary Bill as postmaster, opened at Cloverdale on September 21, 1888. It continued to operate until October 13, 1899. The Belmont-Sodaville stage used Cloverdale in the 1880s, but during this period Cloverdale remained quiet. The only people at the ranch besides the ranch workers were overnight travelers. In the early 1900s an auto stage from Ione to Tonopah ran through Cloverdale. The Cloverdale Ranch is still in operation, and the buildings on the ranch are very old and interesting.

Craig Station
(McKeehan Station)

DIRECTIONS: *From Gabbs, head north on Nevada 361 for 1 mile. Exit right on Nevada 844 and follow it for 4.3 miles. Exit left and follow the road for 2 miles to Craig Station.*

Craig Station was established in the early 1860s as a stop on the Ellsworth road. The station was named for R.V. Craig, who was the first recorder for the nearby Mammoth Mining District. Sam McKeehan, Craig's partner in mining ventures, ran the station. Later the station became a short-lived stop on the Ellsworth-Downieville stage line. The stage line eventually went through to Ione but never achieved real importance. Not much is known about the history of the station, which consisted of a small stone building of modest construction along with a small stable. The stage building, marked by tall trees, still stands at the site. The site is occupied, so be sure to ask permission before wandering.

Deep Well Station
(Black Cabin Well)

DIRECTIONS: *Located 2 miles north of Downeyville.*

Deep Well Station was established as a stop on the stage road from Downeyville to Wellington in the 1870s. When that line stopped running, the station was used mainly as a watering hole. It served an important purpose, however, because it was the diversion point for roads heading to Lodi Valley, Ione Valley, and Grantsville. Only faint foundations of the station remain.

Downeyville
(Downieville)

DIRECTIONS: From Gabbs, take Nevada 361 north for 1 mile. Exit right on Nevada 844 and follow it for ½ mile. Exit left and follow the road for 1 mile to Downeyville.

In May 1877 Patrick, Jeremiah, Demon, and James Downey discovered rich silver and lead deposits just west of Ellsworth. Their discovery lured many people away from Ellsworth, and soon a new camp sprang up at Downeyville. The first business in the town was W. H. Howe's store. A large tent city formed, and by 1878 the town had a population of more than 200. Residents constructed a number of substantial buildings, including stores, saloons, a Wells Fargo office, and stage line offices. A supply stage ran from Luning to Downeyville on a regular basis. T. J. Bell, the Wells Fargo agent, ran a stage line to Wadsworth. Bell served five times in the state assembly and two terms in the senate. He had a large ranch—one of the most extensive in central Nevada—which ran from Gabbs Valley to Reese River Valley. The Yomba Indian Reservation later formed on what had once been the ranch.

In 1879 the town was officially named Downeyville, after the Downey brothers, who had made the first locations. A post office, with George Gates as postmaster, opened there on March 31, 1879. Downeyville thrived for quite awhile, and ore from its mines was shipped to mills in Austin and elsewhere. By 1880 Downeyville was home to the O'Malley Office Building, the Raphael Store, the Spence Saloon, the Curley Saloon, the Travis and Hawk Store and Restaurant, the York and Bickford Boardinghouse, and a school. The Downey Mining Company ran a smelting furnace, an assay office, and a blacksmith shop. The Downey smelter was one of the first true lead smelters in the state. In time, however, the excitement died down, and people left. By 1885 only the Downeys remained.

The main mine in the district was the Downeyville mine, which was more than 500 feet deep. During its production years from 1878 to 1901 the mine is reputed to have produced from $7 million to $12 million worth of silver and lead, although state records list the production value at $600,000. In 1901 the Nevada Company, run by J. Phelps Stokes, took over the mines from the Downeys. The company opened a store and a boardinghouse at Downeyville. They hired about thirty men to work the mines, but paying ore could not be found, and the company left the district in the fall. The post office closed on

October 15, 1901. Soon the only visitors to Downeyville were wisps of wind from Gabbs Valley.

The town enjoyed a brief revival in 1923 and 1924, when Downeyville Nevada Mines, Inc., purchased the old Downeyville mine and began to rehabilitate the mine workings. The company, whose president was F. P. Aylwin of Luning, formed in early 1923. Miners worked the old mine on and off from 1924 to 1927, when the company began operating at a loss. It folded soon thereafter. The district has remained silent since 1927 except for a period in the early 1950s, when J. O. Greenan, Lloyd Mount, and C. A. Carver did some leasing in the area.

The ruins of Downeyville are fairly extensive, although there are no complete buildings left. Over thirty stone ruins are scattered in a small canyon below the remains of the Downeyville mine. The layout of the ruins suggests that Downeyville had two main roads running parallel to each other. Most of the buildings seem to have been small houses for miners. In addition to the stone ruins, faint stone foundations abound along the two streets. The interesting stone ruins of the smelter are directly below the Downeyville mine. The ruins at the mine are among the best-preserved remains at the site and still have the hoist and engine housing. These were used by the Downeyville Nevada Mines company, which is the main reason for their continued existence. The site is extremely desolate but is well worth the trip.

Duluth
(Stratford)

DIRECTIONS: *From Phonolite, continue east for 1 mile to Duluth.*

Duluth, a small mining camp, grew up as a suburb of Phonolite. Prospectors discovered small amounts of ore in the vicinity in July 1906, and a small camp quickly formed east of Phonolite. Within one week 100 people had created the camp of Stratford. Three weeks later Duluth came into existence only a stone's throw away. By October it had absorbed Stratford, and it contained three saloons, a boardinghouse, a lodging house, a butcher shop, a livery stable, a store, an assay office, a barbershop, and two brokerage houses. A semiweekly stage began running to Duluth from Luning, and George Wingfield and Judge L. O. Ray invested in the Duluth mines. By January 1907 miners were working the Last Chance, the Savage, the Jack, the Sarah, the Alpha Omega, the Allis, the Lillie, the Lamont, the Haven, the Sparrow, the Sandstorm, the North Star, and the Jumbo mines, and in February a miners' union formed. Two small mining companies, the Big Henry Gold Mining Company and the Duluth Gold Mining Company, became active in the area.

A post office, with Frederick Hess as postmaster, opened in Duluth on April 27, 1907. A newspaper, the Duluth *Tribune,* was published once a week and distributed in Duluth and nearby Phonolite. The interest in Duluth did not last long, however, and by the end of 1907 most people had moved out. The two mining companies folded in late 1907, and except for some minor claim work no more mining activity took place there. The post office closed on December 14, 1907, and the newspaper also folded. By early 1908 the district was empty. In November 1915 R.E. Bruner, H.W. Bruner, H.M. Wheaton, E.C. Scoy, and William Morse, all former owners of the Big Henry Company, incorporated the Duluth Gold Mining Company, but they had little success. In 1921 the Kansas City–Nevada Consolidated Mines Company, which was prominent in nearby Phonolite and Lodi, worked the Duluth mines for awhile but found little. In 1929 the Golden Eagle Mining and Milling Company reworked the Duluth tunnel and treated its ore at the mill in Phonolite. The company's production was sporadic until it folded ten years later. The property sold at auction in April 1940 to Ole Peterson for $14,000, but he did no work and only sold the remaining equipment.

Ruins at Duluth are scarce. A few dugout cabins remain, but there is nothing else at the site except some faint sunken foundations.

Ellsworth
(Upper Weston) (Summit City) (Mammoth) (Corrine)

DIRECTIONS: From Gabbs, head north on Nevada 361 for 1 mile. Exit right and take Nevada 844 east for 8 miles. Exit left and follow Ellsworth Canyon Road for 6 miles to Ellsworth.

In December 1863 Indians led James Daley and R.V. Craig to a large silver ledge in the hills west of Ione. Sam McKeon and A.T. Hatch organized the Mammoth Mining District the following year. Two small camps soon formed next to the mines in Ellsworth Canyon. Called Ellsworth and Upper Weston (Summit City), they were less than a quarter of a mile apart. Weston, whose townsite was laid out in February 1864, was originally called Mammoth, and Ellsworth was originally its suburb. In 1864 there were fourteen houses and a hotel in Weston, while in Ellsworth there were only a couple of homes. Both towns grew until they were considered one, and the joint settlement was known as Ellsworth. The oldest mine in the district was the Mount Vernon, which James Graham and Manuel San Pedro had located in 1863. General Rosecrans bought the mine in 1864, and during the next two years he spent $70,000 developing it.

A post office, with Daniel Edleman as postmaster, opened in Ellsworth on March 7, 1866. By 1867 more than 400 lodes had been located in the area. The town grew quite slowly through the late 1860s but started booming in 1870 after a steam-powered ten-stamp mill was built there. Soon Ellsworth had a population of more than 200 and included almost thirty buildings. For awhile camels were used to haul ore, and the nearby Camel Springs and Camel Pass were named in their honor. For years afterwards people saw the then-wild camels wandering through the nearby valleys. A number of businesses began to flourish, and a stage and freight line to Wadsworth was established at a shipping cost of $50 per ton.

Principal mines in the district were the Morning Call, the General Lee, the Peoria, the Lisbon, the Silver Wave, and the Mount Vernon. At 180 feet deep, the Mount Vernon was the deepest of the six. Ore from the mines assayed at an average of $100 per ton. Don Manuel San Pedro of Grantsville established a new mine, the Esta Buena, in the early 1880s, and its ore sometimes assayed as high as $1,000 per ton.

Ellsworth's ore became increasingly low grade, and by 1872 the mill only functioned periodically. By 1874 it had closed, and only twenty people were left in Ellsworth. The boom at nearby Downeyville had drained the town. In 1877 the mill was processing old tailings, and before the mill at Grantsville was completed, the Ellsworth mill treated the rich ore from San Pedro's Alexander mine at Grantsville. By 1878 there were just two stores and one saloon left in Ellsworth.

A strange incident occurred in May 1880, when the Ellsworth district mining records were stolen. Six men who were never caught assaulted District Recorder Peter Schiel. A motive for the attack was never established. The town struggled through 1884, but the post office closed on December 29, 1884. After the office closed Ellsworth slowly faded away. In April 1895 Ted Cirac of Grantsville bought the mill and moved it to Union Canyon. In early 1906 Oliver Boyd's discovery just east of Ellsworth led to the quick formation of the camp of Corrine, which soon had a population of thirty. But Boyd's vein proved to be shallow, and Corrine had disappeared by the end of the year. In 1916 the Bluebird Consolidated Mining Company worked the Mexican mine and reworked the old mill dumps, producing $57,000 worth of ore. In 1920 the Return Mining Company opened the Return mine, which was located at 7,600 feet. The company bought the old Liberty mill and moved it to the mine, but the venture proved a failure, and the company folded.

The district remained dead until 1923, when the Tonopah-Brohilco Mines Corporation began working twenty claims in the Mammoth district. The company, incorporated in January 1923 as a reorganization of the Brohilco Silver Corporation, had George Porter as president and Victor Keith as superintendent. Two rich claim groups among the twenty—the Silver Leaf and the Black Reef—produced ore that assayed at an average of $25 per ton. The company did not make the grade, however, and it folded in 1925.

The last company to work the Mammoth district was the New Return Mining Company. It owned a 240-foot mine that yielded ore assaying at $35 per ton and had values in both silver and gold. The company brought in a used twelve-stamp mill and also built a small cyanide concentration mill. The Return Company was active in the district from 1924 to early 1926. After the company left, the district fell silent. It was not until 1928 that mining activity returned, when Donald Benton of Salt Lake City purchased the old Flagstaff mine and a number of adjacent claims. Benton organized the mine and claims into the Early group, and mining operations began, producing ore that assayed at an average of $55 per ton. Benton maintained control of the group until 1933. That year, he sold the property to J. L. Corlett, who had been operating a small mill in lower Ellsworth. Corlett did not have as much luck as Benton did, but he still realized a fair profit. Corlett's departure in 1944 signaled the end of mining in the Ellsworth district. The district's total production value stands at $800,000.

Among the remnants of Ellsworth are five wood buildings in good condition, four of which are still inhabited. About half a mile below these are a number of older stone buildings. They include the remains of the steam-powered stamp mill. The many old stone foundations scattered around the mill ruins make for interesting exploration. The road to Ellsworth, though graded, has some treacherous sand traps.

Frazier Wells

(Frazier Springs)

DIRECTIONS: *From Tonopah, head west on US 95 for 1.3 miles. Exit right onto a tar road, and follow it for 4 miles. Exit left onto a rough dirt road and follow it for 2¾ miles to Frazier Wells.*

Frazier Wells was a small stage stop for travelers heading to the booming mining town of Ray, located a couple miles to the north. In 1901 a small camp of ten formed at the wells, from which its name came, shortly after Judge L. O. Ray made his discoveries there. A few small business establishments housed in tents, including a saloon and a small food store, opened at Frazier Wells. When Ray ceased to exist around 1903 Frazier Wells's residents quickly folded up their tents, and the camp soon became as empty as Ray was. Only a few tall trees mark Frazier Wells, and during hot summers the springs tend to dry up completely. The spot is an excellent resting point, since it offers the only shade for miles around. The road to the site tends to be extremely rough.

Gabbs

(Brucite) (Toiyabe)

DIRECTIONS: *Gabbs is located 65 miles northwest of Tonopah.*

Gabbs Valley was named in honor of William Gabb, a paleontologist with the Whitney Expedition of 1862–1867. While he never visited the area, he analyzed the fossils collected there. In May 1926 Guy Smith and F. S. Stephenson discovered the tungsten that led to the formation of Gabbs. The Nevada Massachusetts Mining Company bought the Betty O'Neal claims soon thereafter, and a camp, named Brucite, formed at the site of the claims. In 1927 Harry Springer discovered brucite in the area and formed the Western Magnesium and Chemical Corporation. Brucite would later become Gabbs's lifeblood.

In 1936 Basic Ores took over brucite production in Gabbs. The company remained active until the 1990s. By 1940 the camp at Brucite contained a ten-room bunkhouse, a mess hall, a kitchen, an office, and numerous cabins. The company had trouble finding enough workers and often hired transients at Luning.

At this time, Gabbs was an empty, dusty place. This changed at the onset of World War II, when magnesium was put on the defense priority list. As a result, Basic Ores was renamed Basic Magnesium (BMI). A special magnesium production plant was designed in a cooperative effort between England and the United States. Unfortunately, final plans for the plant were placed aboard a ship that was torpedoed, and they were lost.

After a lengthy delay, construction of the plant began in December 1941, and the company town of Gabbs was born. Two large dormitories and many homes went up in the town in early 1942. The company strung a $250,000 power line from the town of Millers, and the plant was completed in June 1942. The ore underwent preliminary treatment at Gabbs, and then a convoy of twenty-nine trucks, traveling a combined total of 17,000 miles a day, hauled the ore to the main plant at Henderson for final treatment. Because of the real threat of sabotage, the government hired twelve special police officers to guard the property.

In 1942, in response to increasing demand, a school district was established in Gabbs, and the townspeople built a grammar school and a high school, the latter of which was known as Toiyabe High School until a new high school replaced it in 1955. A jail was also constructed using old jail cells from Belmont. By 1943 Gabbs had a population of 426. In June the Gabbs township was established, and a community house and the Victory Theater, the latter of which later became Gabbs City Hall, opened.

In 1943 BMI was the largest producer of magnesium in the world. During this boom period Gabbs was divided into three distinct and separate sections: North Gabbs, South Gabbs, and Tent City. Tent City was the company town of the Sierra Magnesite Company, which was established in 1941. It consisted of dormitories, a cookhouse, an office, a guest house, and a couple of homes. When the operation shut down in 1948 residents moved most of Tent City's buildings to North Gabbs. Most of North Gabbs's original buildings had come from Silver Peak in 1941. A post office, which was initially called Toiyabe, opened in North Gabbs on December 18, 1942. On June 1, 1943, the office's name was changed to Gabbs. Most of the homes in North Gabbs were constructed from dismantled dormitories. A library formed and is still in operation today. In 1943 South Gabbs contained sixty homes, a swimming pool, a recreation hall, a city park, and tennis courts.

For the next fifty years the size of Gabbs's population held constant, and BMI remained the town's lifeline. Gabbs officially incorporated on March 29, 1955. The new Gabbs High School opened in 1955, and the school's nickname was the Tarantulas. Many short-lived newspapers served the town. These were the *Toiyabe Treasure* (1942), the *Gabbs Gab* (1943–1945), the *Booster News* (1956–1960), the *Gabbs Valley News* (1968), the *Gabbs Valley Enterprise* (1974–1976), and the *Gabbs Independent* (1987–1988).

From 1986 to 1994 the FMC Gold Company ran a large open-pit microscopic gold mine and mill nearby, and most of the workers lived in Gabbs. The town is still alive, and as long as the mines it relies on for survival continue to produce, life there should remain as it has for many years.

Glen Hamilton

DIRECTIONS: *From Ione, head north on old Nevada 21 for 9 miles. Exit left and follow the road for ¼ mile. Exit left again and follow the road for 2 miles to Glen Hamilton.*

Glen Hamilton was a small stage stop on the Ione–Austin stage run. There was a post office at Glen Hamilton from May 18, 1866, to October 15, 1866. Andrew Hawk was postmaster. Only a small log structure and corral were built at the site. Few people ever lived at Glen Hamilton, and why it warranted a post office is a mystery. Today nothing at all remains in the area, and the site location is approximate.

Globe Mining District
(Kelly's Well) (Nickolay Camp)

DIRECTIONS: *The Globe Mining District is located 6 miles south of Gabbs, ½ mile east of Nevada 361.*

Miners organized the Globe Mining District in 1883, after James Sullivan discovered the Globe ledge and began developing the Sullivan mine. Patrick Downey was elected district recorder. Interest in the district increased when Richard Kelly discovered the Sunrise mine there in 1886 and when Harry Somerville and Dr. Laws began working five other mines in the area in 1887. Prospectors built a small mill at Kelly's Well and a pipeline to the Sullivan mine. But the mill was a failure, and the miners abandoned it the following year.

During the mid- to late 1880s and 1890s a small camp developed near the mines. It consisted of miners' cabins and had only about ten residents. In 1892 the Downey brothers took over the Sullivan mine. The Nevada Company, which J. Phelps Stokes ran, bought the mine in 1905 but gave up on it the following year. The last bout of activity took place in the district in 1909, when the Rattler Mining Company opened the Rattler and Mazie M. mines, but the mines produced little before the company folded in 1911.

Today, mine dumps and old cabins are scattered around the district, and ruins of the mill are at Kelly's Well. During the 1880s Kelly's Well served as a water stop on the Luning-to-Downeyville road. Later, the well fueled the ill-fated attempt to use traction engines to pull ore wagons.

Golden
(East Golden)

DIRECTIONS: From Cloverdale, continue north for 1 mile. Take the right fork in the road and follow it for 4½ miles. Exit left and follow the road for 3 miles to Golden.

Indians struck gold at Golden in 1902, but no real activity took place there until January 1906, when a rich free-gold vein was discovered. A small camp formed at the site, and it appeared as if the gold would last awhile. A post office, with William Coppernoll as postmaster, opened on February 26, 1906, and served the fifty people living in the camp. The ore was not of a consistently high grade, however, and when prospectors discovered large deposits at Manhattan and Round Mountain, most of the people in Golden moved to these new strikes. The final blow to Golden came in the form of the San Francisco earthquake, since the town's financial backing came from that city. When Golden's backers pulled out, the camp died.

A small number of people stayed behind at Golden and continued to work both the small mines and the Cloverdale placers located just east of the camp. The post office closed on December 14, 1907, but the camp continued to struggle along for a number of years before finally giving out. In 1910 there were still thirty people living in Golden, but by 1913 they had all left.

There has been a great deal of activity in the Golden area since 1913. In October 1920 Barney Francisco staked ten claims just north of the townsite, and a shaft sunk on one of the claims was called the East Golden mine. Francisco built a homemade tube mill at the mine, but it was a failure. He leased the mine in 1941, and the leasers built a Huntington mill, complete with crushing plates and a concentrating table. Assays from the mine ran from $20 to $30 per ton. During the early 1920s miners worked both the Davis and the Humphrey properties.

The Cloverdale placers were reworked in 1931. A company sank twenty-six shallow shafts, ranging from twenty feet to fifty feet, in an effort to locate substantial deposits, but the search was fruitless. In 1939 Wadley and Hunt bought the West Golden claims, owned by Ben Minton and Roland Tidwell, for $55,000. The claims had produced $10,000 worth of ore since 1932, but the two men had very little success and gave up in 1941. In the early 1950s miners once again worked the West Golden claims, but they recorded no production.

Nothing but a partially collapsed wooden cabin marks Golden today. The head frame and another cabin still remain at the East Golden mine, but the shaft has completely collapsed. Only scattered shallow shafts mark the site of the Cloverdale placers.

Goldyke

(Tom Burns Camp)

DIRECTIONS: From Gabbs, head south on Nevada 361 for 2 miles. Exit left and follow the road for 10 miles to Goldyke.

Goldyke, a small mining camp, sprang up after discoveries in nearby Atwood set prospectors to searching the mountains and canyons around that booming camp. These prospectors discovered gold a mile southeast of Atwood in 1905, and Goldyke quickly formed. A townsite was platted in August 1905, and one of the first businesses in the town was H. G. Jackman's saloon and gambling hall. In the fall of 1905 the Gold Reef Mining Company completed construction of a mill called the Richardson. A post office, with Phillip Somerville as postmaster, opened on January 9, 1906. Soon afterwards a newspaper, the *Goldyke Daily Sun,* began publication. In 1907 an auto stage, run by Dee Jones, ran from Goldyke to Luning and was soon in full swing. Goldyke peaked in 1907. Its businesses included the Pioneer Cash Store, owned by W. D. Jones; the Goldyke Barber Shop, owned by Harry Wheeler; and a red-light district that was home to more than a dozen prostitutes. The townspeople organized a baseball team and played a team from Atwood on Sundays. Because there were a few local deaths, a small cemetery formed between the two towns.

The veins at Goldyke were quite shallow, and the town faded as quickly as the gold did. The post office closed on October 15, 1910, and Goldyke followed suit. Some scant activity took place there in the 1940s, when Tom Burns worked the Jim group of claims, located one-fourth of a mile from Goldyke. Burns had arrived at Goldyke in 1906 and for years after its collapse was the only resident. Miners erected a few buildings at the claims, but this bout of activity also quickly petered out. Soon Burns was once again the only person in the district. He was still prospecting the area when he died in 1951.

The remains at Goldyke are few. They consist of just one partially collapsed building and the ruins of the five-stamp mill. Other than that, only scattered debris marks the site.

Granite

(Douganville)

DIRECTIONS: Granite is located 3 miles south of Lodi.

Jack Hughes and J. H. Hatterly made the first discoveries in the Granite district in July 1922. Initial assays were more than $900 a ton, and the owners of the Betty O'Neal mine (in Lander County) bought the mine for

$150,000. At around the same time the Brohilco Silver Corporation opened the Silver Leaf mine, but the ore pockets in both mines had been mined out by 1923.

It wasn't until the 1940s, when Albert Brown, William Sovy, and John Sutherland discovered tungsten, that Granite became active again. The Gabbs Exploration Company formed in 1945 and purchased all the claims in the district for $100,000. The company built a mill at Gabbs in 1951. A small company town, called Douganville after James Dougan, came into existence at the Victory Tungsten mine in 1955. The company employed thirty-five men, and operations continued until 1958. There are still extensive remains at the Victory Tungsten mine.

Grantsville

DIRECTIONS: *From Ione, take old Nevada 91 south for 6.4 miles. Exit left and follow the road for 1½ miles. At four corners, take a right and follow the road for 4 miles to Grantsville.*

P.A. Havens, who was already well known for his discoveries around the Ione area, discovered gold in Grantsville Canyon in 1863. Havens immediately organized the site into a mining district and laid out the ground-work for a good-sized town, which was named after Ulysses S. Grant. Havens sold lots in the town for between $50 and $500. Soon about fifty people were residing in the picturesque canyon. But Havens's rich claim quickly dwindled, and the growing camp became a ghost town for the next ten years.

The camp was reborn in September 1877 when the Alexander Company came to Grantsville at the suggestion of Manuel San Pedro, an original inhabitant of the town. The company bought a number of claims in the canyon; built a twenty-stamp mill, which it enlarged to forty stamps three years later; and hired thirty men. The mill started operating in November 1878. The company made some very good discoveries, and the population of Grantsville soon swelled to almost 1,000. By the end of 1878 businesses in Grantsville included the Humphrey Hotel, the Grantsville Drugstore, the Grantsville Laundry, the Bonanza House, the Alexander Market, G.B. Smith Livery, the Cirac French Bakery, and Howe General Merchandise. Saloons included the Star, the Alexander, the Mexican Union, and the Exchange. T.J. Bell also started a weekly stage from Belmont.

D.L. Sayre ran Grantsville's first newspaper, the *Grantsville Sun,* which "dawned" on October 19, 1878. The paper did not last long, folding in June 1879. Its successor was the *Grantsville Bonanza.* Andrew Maute and Samuel Donald were its publishers. It started publication on December 11, 1880.

Maute also published the *Belmont Courier.* The paper struggled through 1884 but finally folded due to lack of public interest.

A post office, with George Healy as postmaster, opened in Grantsville on February 3, 1879, and mail arrived three times a week. Three stage lines—to Eureka, Wadsworth, and Austin—were set up, and by 1881 the town had over forty business establishments, including ten stores, five saloons, two assay offices, an express office, and a bank. The Odd Fellows built a school that housed sixty students, and the brick building is one of the few structures still standing in Grantsville.

There were fourteen major silver mines in Grantsville Canyon, the best producer of which was the Alexander mine, with a 1,200-foot shaft and a 500-foot incline shaft. The Brooklyn mine, which was discovered in June 1880, was another steady producer for the Alexander Company. The canyon mines produced more than $1 million in gold and silver before 1885, when the district was temporarily abandoned.

Grantsville was the scene of a number of murders. The first of these occurred on August 10, 1880, when Thomas Burn shot and killed Thomas Mack while he slept. The men had been drinking together, and no motive for the killing was ever established. The second and third murders took place in March 1881 when Mattias Salmon shot and killed a popular Grantsville resident, S. Merrill, for no apparent reason. Salmon was thrown into the Grantsville jail, which was an abandoned mine. An angry group of Grantsville citizens broke in, dragged the luckless Salmon to the stamp mill, and

The one-room school-house at Grantsville struggles to stand. (Shawn Hall collection, Nevada State Museum)

lynched him, hanging from the crossbeams. The death was later ruled a justifiable homicide. Needless to say, these killings got the Grantsville cemetery off to a good start.

Grantsville faded slowly in the 1880s. In November 1880 the Alexander Company, whose mines had produced $1.25 million worth of ore, was forced to stop all operations due to litigation. In 1881 San Pedro resigned as superintendent to concentrate on his interests in the Gold Park district. Finally, with the litigation mired in the courts, F. S. Van Zandt bought the company for $675,000, but he ended up forfeiting, and the county was stuck with the property.

By 1884 Grantsville's population had diminished to 400. The camp kept functioning, but barely. By 1886 just fifty people were left. John Phillips leased the Alexander property and ran the mill on old Alexander and Brooklyn tailings. In June 1888 the Hornsilver Mining Company bought the Alexander property from the county. Thomas Mitchell, who had been superintendent at mines in Tybo, Reveille, and Union, ran the company. The company hired twenty men, and the operation enjoyed limited success until Mitchell's death in 1897. Without his guidance and knowledge, the company soon folded.

Grantsville was stubborn. Every time the town seemed to be on its last leg, a new revival would keep it going a little longer. The Grantsville post office closed on October 31. By then, only a handful of people remained in the dying town. In January 1907 Charles Houchel bought all the Grantsville mines for $750,000 but didn't last until the end of the year. In addition, Bob Hanchett salted a mine, in other words, he planted gold in the mine to make the mine appear valuable, and his rich "strike" attracted forty miners back to Grantsville. Ed Kiefer opened a saloon and a prostitution hall, but as soon as everyone realized the gold find was a scam, they all left. W.J. Webster reopened the Alexander mine in September 1916 but after having limited success gave up in 1918.

A small-scale revival between 1921 and 1923 brought some activity back to Grantsville. The Webster Mines Corporation—based in Wilmington, Delaware—bought the old Alexander mine, renamed it the Webster mine, and hired forty men. The company, run by Elmer Bray and Jay Carpenter, soon abandoned its efforts, and the district remained silent for four years. The company kept control of the property and leased the mine to William Hooten and associates in 1927 and then to the Stabler family of Los Angeles in 1928. The company remodeled the stamp mill, changing it to ten stamps and adding a rod mill and two flotation cells of fifty-ton capacity, driven by a distillate engine. The finished concentrate contained 200 ounces of silver and was almost 30 percent lead.

The district was quiet until 1939, when a man named Barrows purchased the Silver Palace mines and began operations. He built a fifty-ton flotation mill near the mines in the fall of 1939. The mill produced $100,000 for Barrows before he closed it in 1940. The last bout of mining activity in the Grantsville district took place from 1945 to 1947, when the Alexander and Brooklyn Mines Company, of which E.D. Feldman was president and which employed twenty-seven workers, reworked some mines and obtained $50,000 worth of lead concentrates for the war effort. The company left the district in 1947 after its owners were indicted on fraud charges. Grantsville was abandoned once more, this time becoming a permanent entry on the ghost-town register.

The ruins of Grantsville, which are some of the best in Nye County, range from the solid brick schoolhouse to small stone cabins. The schoolhouse still stands, although it has been heavily vandalized. The main streets are easy to locate, because long rows of foundations mark the once-booming business area. Only one building, which appears to have been a blacksmith shop, remains in this district. A few more recently built wood cabins are north of what used to be the center of Grantsville. A large spring, coming from a pipe, runs freely through the ruins. The quality of the water for drinking is questionable, for it has a high mineral content. A cemetery is located on a hill near the old townsite.

The ruins of the stamp mill are just below the brick schoolhouse, and a few buildings remain in front of the mill. The brick building served as an assay office. An old bulldozer, left from the World War II revival, is slowly rusting in front of the mill. An interesting aspect of the author's exploration of Grantsville was the discovery of an old safe, with its door blown off, about half a mile below the mill. There was no clue as to the origin of the safe.

Plan to spend a day in Grantsville. The ruins are so extensive and intriguing that it would be a crime to hurry through them. By poking around the many foundations, it is fairly easy to determine the sort of establishment that was once housed in each. The author rates this ghost town as one of Nye County's best.

Green Springs
(Iron Bed Springs)

DIRECTIONS: *Green Springs is located 4 miles east of Downeyville.*

Green Springs was a short-lived mining camp that sprang up during the summer of 1907 after Clinton Beard discovered a small deposit of gold in March. A tent camp of fifty inhabitants with residences, saloons, and restaurants formed, but Beard's claims were the only ones that had any value, and everyone but him left by the spring of 1908. Surprisingly, however, the camp was awarded a post office on July 23, 1907. The office closed on March 19, 1908. Beard at one time refused an offer of $50,000 for his claims. The vein subsequently disappeared, and he later died penniless. Only some small prospecting dumps mark Green Springs.

Idlewild

DIRECTIONS: *Idlewild is located 4 miles north of Ione in Idlewild Canyon.*

Idlewild Canyon was the site of a small offshoot camp of nearby Midas. Only a handful of prospectors ever resided there, and no actual mining took place at the site. Little is known about the history of the settlement. The Boy Scouts of America have recently used the Idlewild Canyon area as one of their outing places. A large, old log structure whose purpose is unknown remains at the site, but nothing else is left. Fresh spring water is available.

Ione

(Ione City) (Midas)

DIRECTIONS: *From Gabbs, head north on Nevada 361 for 1 mile to Nevada 844. Take a right and follow the road for 22.1 miles to Ione.*

Ione formed in November 1863 soon after P.A. Havens made initial silver discoveries in the Union district. At first, the camp had only a handful of residents, but within a few months there were over fifty buildings and more people. By January 1864 the residents of Ione were demanding that a new county be formed because of the richness of the Ione area. The territorial government consented in January 1864, and an official county government was organized on April 26. The town was allotted $800 with which to build a county courthouse, and present-day residents say it was housed in a small wood cabin that still stands. By the spring of 1864 the town had a post office, more than 100 buildings, and a population of 600. A board of county commissioners formed in Ione in April 1864.

Henry DeGrout and Joseph Eckley organized two newspapers that began publication in Ione in 1864. The *Nye County News* printed its first issue on June 25. It ran into trouble and folded after about a month but was reborn on July 1, 1865, with Eckley now the sole proprietor. Despite glowing promises, the paper encountered further difficulties and finally ceased operations for good in May 1867. The second paper, the *Advertiser,* was extremely short-lived. It was first published on September 17, 1864, but expired on October 29.

By the end of 1864 businesses in Ione included the Griffin Restaurant; the Eclipse Stable, owned by Thomas Morgan; the Empire Market, owned by Peter Schiel; the Fashion Saloon; the Ione House, owned by Mrs. Dunbar; the Bank Exchange Restaurant; the Silvester Store; the Ione Drugstore, owned by B.W. Crowell; the Pioneer Store, owned by Sam Hyman; the Ione Meat Market, owned by John Burcham; and the Ione Store, owned by D.C. Clark. Thomas Morgan ran the Austin–Ione Stage, and Henry Whittell ran the West Gate and Ione Toll Road. To help promote the town, John Sharp established the Ione City Town Agency.

Most of the mining in the Union district was located miles from Ione, but a five-stamp mill, the Pioneer, was in Ione. The mill never turned a real profit, and it closed in 1866. New strikes in Belmont in 1867 lured away many of Ione's residents. The county seat's relocation to Belmont in February 1867 dealt the town a bitter blow. Ione began a quick downhill slide, and its population had sunk to 175 by 1868. The town experienced many ups and downs during the next decade. In the 1870s Ione had a few good years of production, but Belmont remained the focus of attention, and Ione's population continued to shrink. During the mid-1870s seventy men were working

On Independence Day
in 1897, Ione included
a store, formerly at
Berlin, and a number of
false-fronted buildings,
of which only a few
remain today. (Nevada
Historical Society)

at the Storm King mine and another twenty at the Stonewall mine, providing most of Ione's business. By 1880 the town's population stood at twenty-five. Only a few business establishments still functioned. On April 8, 1882, the townspeople changed the post office's name to Midas, possibly in an attempt to change the town's image.

A severe fire struck Ione in September 1887 and destroyed the W.S. Gage storehouse, three homes, a saloon, and a livery stable, for a total loss of $10,000. Ione received a boost in 1896 when a new ten-stamp mill was completed. The Ione Gold Mining Company owned the mill, which E.W. Brinell managed and which treated ore mainly from the Berlin mine. The completion of the mill helped revive Ione, and the population rose to seventy. In 1897 the Nevada Company, run by J. Phelps Stokes, arrived in the district and bought most of the mines and mills in the area. One partner, local stage magnate T.J. Bell, sold out in 1898 and bought the New Pass mines (in Lander County).

By 1898 there were 175 men working at Ione. In March the Nevada Company bought out the Ione Gold company for $150,000. The company also bought the Citizen's mill in Austin and moved it to Ione, but a dramatic drop in the price of silver forced the company to curtail operations in July. The revival was over, and the post office closed on January 15, 1903. When it reopened on July 16, 1912, it was called Ione once again.

A mining revival woke Ione when the Midas Gold Mining and Milling Company, the Manchester Mining Company, and the Tonopah Ione Mining, Milling, and Leasing Company all began operations. A telephone line to Austin was completed, and Ione's population rose to 100. The revival came to an end in 1914, and on April 30 the office once again closed, this time moving to Berlin. The Berlin office closed in December 1918, and the Ione office reopened on December 18 of that year.

The new mining revival centered on mercury. In the initial excitement over gold and silver, the rust-colored cinnabar ore had been overlooked. The Mercury Mining Company, which had its principal properties in nearby Shamrock Canyon, mined most of the mercury. The company produced 11,000 flasks of it, each weighing seventy-five pounds. A large mill began operations on the edge of Ione but was never used consistently.

Mercury mining continued into the 1930s. After that bout of activity came to an end, mining at Ione was sporadic. But the town clung to life. In 1945 its population still stood at forty-five, but when the post office closed on April 30, 1959, only a handful of residents were left. Estimates of the total value of production for Ione and the immediate area range from $500,000 to $1 million.

The mill built for mercury treatment was torn down in 1950, and only its foundations remain. Today the little town quietly sleeps. Ione still retains its pioneer flavor and has an abundance of interesting old buildings. Among the better ones are the schoolhouse and a few false-front stores. A number of very old stone cabins are also left. Gas and limited grocery supplies are available at Ione's one remaining store. This quaint town is a definite must for any ghost town buff.

Jackson Mining District
(Gold Park) (Barnes Park) (North Union)

DIRECTIONS: *From Ione, head northwest out of town for 12 miles toward Panelas. Exit right (sharp) and follow the road for 10½ miles. Exit right again and follow the road for 1½ miles. Exit right once again and follow the road for ½ mile. Turn left and follow the road for another ½ mile. Take a right at the fork and follow the road for ½ mile to the Jackson Mining District.*

The Jackson Mining District is on the Lander–Nye County border. It is included in this book because its major mines were located in Nye County.

Prospector Thomas Barnes made the first discovery of ore in the Gold Park Basin in 1864. Shortly afterwards the North Union mining district formed. By 1865 active mines in the district were the Summit, the Nevada, the Clad, the Monarch, the Golconda, and the Mary Gray. Ore values were low, however, and by 1867 most activity had ended. In 1878 the district was reorganized and was renamed the Jackson Mining District. By that year there were three principal mines and seventeen other sites in the area. The three principal mines (the San Francisco, the Arctic, and the North Star) all had vertical shafts of just over fifty feet.

In 1880 another group of prospectors, led by Frank Bradley, found more ore deposits in the district. Soon operations in the Gold Park Basin picked up. The Star of the West and the Arctic were the most productive mines in the area. A ten-stamp mill was completed in the fall of 1881, but it turned out to be inefficient, as it cost $300 to produce $100 worth of ore. By 1883 the only people left in the district were some prospectors. In 1893 the district had three large claims and a mill, all of which the Nevada Mining Company then purchased. In 1896 the company built a new stamp mill, which continued to operate until the teens. In 1898 J. Phelps Stokes's Nevada company sold the property to the Gold Park Mining Company, which operated until 1919. In 1919 Robert Todd, who organized the Star of the West Mining Company, bought the property. Todd built a new fifty-ton amalgamation and concentration mill, which was completed in 1921.

The Star of the West company was originally incorporated in Delaware in 1919 with $1 million in capital. By the end of 1921 it owned fifteen mining claims and three mills in Gold Park Basin. It also owned over 4,000 feet of workings and the fifty-ton mill. Activity in the basin ceased shortly after 1921, however, when the quality of the gold and silver ore declined. From 1939 to 1942 Albin Nelson worked the Peterson, the War Eagle, the Lookout, the Boyd, and the Gooley mines, but he had only limited success. When he left the Jackson Mining District was abandoned for good. The value of the district's total production easily exceeded $1 million.

The district was in an excellent spot, with an abundance of water and wood in the surrounding mountains. It was known not only for its silver and gold but also for its semiprecious stones including geodes, agates, and other gems. No buildings remain in the basin, but evidence of activity is still visible. Tailing piles can be found throughout the area, and the mill foundations make for interesting exploration.

Jett

(Argentore) (Silver Point) (Garrard) (Davenport)
(Birmingham) (Gibraltar)

DIRECTIONS: *From Round Mountain, head west on Nevada 378 for 2.7 miles. At the junction with Nevada 376, take a left and follow the road for 1½ miles. Exit right and follow the road for 6½ miles to Jett, located in scenic Jett Canyon.*

John Davenport and John Jett discovered the Jett Mining District, whose original name was Garrard, in 1875. The district officially formed the following year. A flurry of activity took place in 1876. Mines started were the Idlewild, owned by Davenport, Jett, E. E. Shumway, and A. E. Ashburn; the Centennial, owned by the Centennial Mining Company; the 76, owned by George Nicholl and A. T. Vollmer; the Mammoth, owned by Eggleston, Thomas, and Hancock; the Mountain Boy, owned by Davenport and James Griffin; the 4th of July, owned by Griffin; the Boston, owned by C. B. Snow; the Fairy, owned by Colonel Hollis; and the French Republic, owned by R. Lazard. The district's three major mines were the Centennial, the 76, and the Idlewild. The Centennial was easily the richest of all, with its ore assaying between $100 and $300 a ton. The other two mines had relatively low-grade ore that contained no gold and only small amounts of silver.

In February 1877 a small camp formed and was officially named Davenport at a town meeting. Twenty miners were living in the camp, and Ashburn and Shumway, who also ran a large store in Jefferson, built a thirty-foot hall that served as the community center. Jett's first business was Watkins's and Bollen's butcher shop. In 1877 the Idlewild Company built a smelting furnace, but it was a failure. The small town suffered a blow in December 1878, when John Jett suddenly died at Moores Creek Station. He was buried in Jefferson, and the whole camp of Davenport attended the funeral. As a result of Jett's death, the camp was renamed Jett at a town meeting held after the funeral. In December 1879 the Jett Consolidated Silver Mining Company formed, and John Counter was named superintendent. He was also the superintendent of the San Antonio Silver Mining Company.

In response to the activity that took place in 1880 a post office, with Thomas Warburton as postmaster, opened in Jett on March 16, 1880. The office was in Warburton's store, the first store in town. Charles Engstrom opened a saloon, which proved very popular. The ore mined at Jett was shipped north to Eureka for smelting. Unfortunately, the ore's quality made working it properly impossible. Supplies were freighted in from Austin, sixty miles to the north, at a cost of $30 a ton. Activity slowed in the district in 1881, and on April 21 of that year the post office shut down. From 1881 to 1883 the only active mine in Jett was the Silver Point, and the camp was once again renamed, this time after the Silver Point mine. A New York firm that was working the new Senator mine started a brief revival. The post office reopened on June 6, 1890, but the excitement faded as quickly as the ore did, and the office closed for good on March 25, 1891. By April Jett had been abandoned. Tony Fisherman made a new strike in May 1912, and interest in Jett Canyon revived. John Reeves staked out a townsite called Birmingham. George Wingfield took an option on the mines the following year, but he found little of value, and his departure dashed any hopes of a sustained revival.

The district remained dormant until May 1919, when a rich vein was discovered near the mouth of Jett Canyon. A new settlement, Argentore, also called Gibraltar, formed at the mouth of the canyon, two miles from the original site. Sam Forman and Harry Stimler discovered the new vein, named the Gibraltar. The two men quickly sold out to the New York–based Gibraltar Silver Hill Mining Company in November. The vein contained fairly high quantities of silver, lead, and zinc, along with antimony, and the new discoveries gave rise to a settlement with a huge boardinghouse and a number of cabins.

The vein was relatively short, and the company left in late 1921. In 1922 all its equipment was sold and moved to Silverton. Harry Stimler then took over

Ruins of an unsuccessful mill in Jett Canyon built in the 1880s. Note the waterwheel on the right side. (Bruce Larson collection, Central Nevada Historical Society)

Only structure left in Jett Canyon. (Shawn Hall collection, Central Nevada Museum)

most of the mines and kept the camp alive. In October 1922 Stimler, Nick Abelman, and Jack Jordan sold the Gibraltar North Extension for $20,000, and the new owners formed the New Gibraltar Mines Company. But activity slowly ground to a halt, and by 1925 the district was completely abandoned. Jett's last burst of activity took place in the early 1950s when Louis Cirac, Bob Marker, George Barra, and T. J. Nicely worked some tungsten claims but were unsuccessful. The Jett Canyon mines produced a total of $200,000 worth of ore.

At Jett, a cabin and the workings of the Centennial mine still remain amid breathtaking scenery. The canyon is beautiful and cool. It is hard to believe that just a few miles away is a bleak desert valley. Nothing remains of the settlement at the mouth of Jett Canyon, which is marked only by some faint foundations and wood scraps.

Knickerbocker

DIRECTIONS: *From Ione, head south on Nevada 844 for 1¾ miles. Exit left and follow the road for 1½ miles to Knickerbocker.*

Knickerbocker was a small milling camp that formed around the impressive Knickerbocker mill built in the summer of 1865 with money provided by Eastern backers. Construction of the mill, which was completed in early 1866, cost more than $130,000. The Knickerbocker Nevada Mill and

Mining Company, which also owned the Olive, the Phoenix, and the Lockwood mines, owned the mill. Knickerbocker Canyon was originally called Veatch Canyon, after George Veatch, superintendent of a mining company at Union. The mill, consisting of twenty stamps and six roasting furnaces, operated from 1866 to 1875. It was the principal processor of Grantsville's ore until a mill was built there in 1869. During the mill's years of operation as many as twenty-five people resided nearby.

After the mill closed in 1875 the canyon quickly emptied. In October 1877 the Ural Mining Company restarted the mill to work ore from Grantsville, but once the Alexander mill was completed there in 1878 Knickerbocker closed down again. In 1887 Thomas Mitchell, the mill's owner, reopened and repaired it to treat ore from the Cincinnati Mining Company, which was running the Berlin mine. When the company ceased to operate in 1889 the Knickerbocker mill closed once more. In 1896 W. S. Gage bought it to treat ore from his many mines, but he died in 1897, and the mill closed. No other activity ever took place in Knickerbocker. In 1898 J. Phelps Stokes's Nevada Company purchased the mill and its contents for $23,000, and all the machinery was moved to booming Berlin to be used in a new thirty-stamp mill.

Today the beautiful rock ruins of the mill dominate the old camp. The walls reveal that the mill was extremely large and elegant. A few small stone cabins remain in the canyon. The intricate workmanship that went into the construction of both the mill and the cabins is something rarely seen in

Ruins of the Knickerbocker Mill. (Shawn Hall collection, Nevada State Museum)

structures erected in a mining camp. The site is rewarding, but be on the alert for rattlesnakes, which abound in the ruins.

Liberty

DIRECTIONS: From Tonopah, head west on US 95 for 1.1 miles. Exit right on old Nevada 89 for 12.3 miles. Exit right again and follow the road for 3½ miles. Exit right once more and follow the road for 3 miles to Liberty.

Liberty was a mine rather than a mining camp, although a small group of workers did reside at the mine. Prospectors discovered the Liberty mine in 1867 and built a small mill next to it in 1868. Miners worked the mine until 1874. By the time it closed, the shaft was more than 700 feet deep and had over 10,000 feet of lateral workings. During its seven years of activity, the Liberty mine produced ore worth close to $112,000 in total. The ore was primarily silver, with small traces of gold. In 1879 the San Antonio Mining Company, run by William O'McDowell, purchased and consolidated the mines at Liberty and nearby Potomac. O'McDowell named John Counter, long a prospector in the area, superintendent. The mines involved in the consolidation were the Liberty, the Potomac, the Springfield, the Twilight, the Rigby, the Vulcan, the Las Animas, the Aerolite, the Imperial, the Ivy Green, and the Phoenix. The effort to consolidate failed, and it wasn't until 1883 that Liberty became active again. In 1883 the Arlington Leaching Works began reworking the old Liberty mine dumps, but they gave up in 1885.

The mine remained quiet for many years. Eventually the Tonopah Liberty Mining Company purchased the property and reopened the mine in 1904. Enough activity took place that in 1905 the Liberty–Silver Peak Railroad was planned, but it was never built. In April 1906 W.W. Watterson sold nine nearby claims to the Tonopah Wonder Mining Company for $8,500 and a large block of stock. Soon afterwards the Cramp brothers—Eastern ship-builders—acquired the majority of the stock of all the companies working the district. The Tonopah Liberty company's 125-ton mill began operation in October 1910. By January 1911 the company was shipping $25,000 worth of ore a month. The company operated the mine until 1912 and produced almost $500,000 worth of silver and lead. Declining ore values cut heavily into production profits, and the company sold the mine along with nearby claims to the Liberty Group.

The Liberty Group, headed by James Lindsay of Philadelphia, discovered a new ore body that assayed sixty ounces of silver with small traces of lead. This ore body turned out to be quite small, and all activity ended. In 1915 W.H. Aubrey made the first molybdenum discovery, but at the time there was no market for the mineral. The North Tonopah King Mining Company purchased all claims in the district in March 1926 and reopened many of the mines, but it didn't find enough ore and gave up in 1928. In the 1930s a number of leasers tried to make a living from Liberty's mines but had little luck. J.W. Van Winkle and Antone Johnson made some shipments in 1935 and 1936. In 1943 the Metals Reserve Corporation leased claims owned by Lee Hand and Clarence and Mary Hall and began mining molybdenum. Once the war was over demand dropped, however, and the company left. The Halls continued to work the molybdenum mine through the mid-1950s.

No other activity has taken place at the site since then, although the Anaconda Mining Company began an open-pit operation about a mile north of the Liberty mine in the late 1970s. However, after a few years of digging the pit to reach the molybdenum deposit, the price of the mineral fell, and actual production never took place. The operation was to have employed 600 men, and extensive housing constructed on the outskirts of Tonopah has basically remained unused. Only the stone foundations of the mill and the collapsing shaft mark the site of Liberty today.

Lodi
(Bob) (Marble) (Lodi Tanks)

DIRECTIONS: *From Gabbs, head north on Nevada 361 for 3½ miles. Exit right on Lodi Valley Road and follow it for 7 miles. At Lodi Tanks, exit left and follow the road for 2 miles to Lodi.*

The Lodi district formed around the Illinois mine, which Alfred Welsh and John Kirkpatrick discovered in 1874. Mining had already been taking place in the district, across the valley at Marble Falls, since April 1868, when A.L. Hatch discovered the Marble Falls ledge. During the next two years, miners worked the Montana, the Marble Bluff, the Marble Point, the Piute, and the Union mines. But by the end of 1870 the mines had been abandoned.

The Lodi district was officially organized on May 14, 1875. Prospectors sank a 1,000-foot shaft at the Illinois mine, and William Raymond, head of the Raymond and Ely Mining Company of Pioche, erected a ten-ton smelter. In 1877 the Argent Mining Company formed and began working the Los Angeles, the Sand Mound, and the Downey mines. By 1878 Lodi had a population of more than 100. Businesses included a store, a blacksmith shop, a boardinghouse, a saloon, and a number of cabins. A freight line began running from Lodi to Wadsworth, 100 miles to the northwest. Shipping costs were $45 a ton. By 1879 the Argent company was in deep financial trouble. While all its personal property was sold at auction, the company was able to keep control of its mines until 1881, when they too were sold to satisfy a $43,000 judgement. The Illinois mine closed in 1880 after producing $400,000 worth of ore. The silver ore that had been removed assayed as high as $500 a ton, but the deposits were not consistent. Prospectors made more than twenty-five claims in the Lodi district by the end of 1881, but only six people remained to work them. They were Alfred Welsh, J.H. Massey, A. LeBeau, Archie Farrington, James Graham, and John McComb. In July 1887 Welsh, who was still working the district, purchased all the holdings of the old Argent Company for $67,190. However, he still concentrated on his Illinois mine, the only proven producer of the district. His life ended abruptly in November 1891, when his brother Manuel shot and killed him. Manuel claimed that Alfred owed him $16,000 and that when he confronted Alfred, Alfred went for his shotgun, giving Manuel no choice but to shoot him. However, witnesses who found the shotgun said that it was across the room, unloaded, and covered in dust. Despite this, Manuel was found innocent by reason of self-defense. The Lodi property stood idle, and in 1893 the county finally sold it to Timothy Phelps. Little was done with it, however, and, for the most part, the district was abandoned.

The district remained very quiet until 1905, when new discoveries were made. Three separate settlements developed. The one at the Illinois mine was called Marble. Just to the east was Bob. A new camp at Lodi Tanks was selected as the new Lodi townsite. A post office, with Charles Grill as postmaster, opened at Marble on March 2, 1906.

Beautiful panorama of Lodi Valley from the Lodi townsite. Phonolite is in the mountains to the left and Lodi Wells at the center. (Shawn Hall collection, Nevada State Museum)

Preserving the Glory Days

Main shaft of the Illinois Mine, complete with steam boiler for hoist. (Shawn Hall collection, Nevada State Museum)

The Lodi Mines Company (owned by Chauncey Burt), which had its mine offices in Luning, formed and used steam tractors to haul ore to Luning for a while. The company owned twenty-two claims in the district, including the Illinois mine. Miners worked the Illinois to a depth of 1,060 feet, with more than 4,000 feet of drift work. The company leased some of the claims to companies such as Rural Mines, Inc., United Lodi Mines Company, and Illinois Nevada Mines Corporation. All told, the company employed eighty men.

The townsite at Lodi Tanks, called Lodivale, developed rapidly during the first few years of the revival. Water was plentiful. Soon residents had constructed a number of saloons, a few mercantile stores, and some corrals. The Lodivale post office opened on July 23, 1909, but closed on August 15, 1910. By 1909 Bob included a hotel, restaurants, saloons, a red-light district, and a school. The Illinois mine was beginning to be a very steady producer. In June 1909 a new 100-ton smelter was built to help handle the large amounts of ore being produced. However, the smelter only operated for a little more than a month before it was abandoned. As time went by, the cost of shipping ore bankrupted the company, which was $150,000 in debt. All its property sold at auction for $30,000 in May 1911. The Adaven Mining and Smelting Company, in which George Wingfield was a major stockholder, then took over the mine. Mining operations faltered in 1914 when water began to fill the mine. Adaven sold out in 1915 to the White Pine Mining Company, but it was unsuccessful. Eventually Chauncey Burt once again had all his property back. The towns struggled to survive, but the Marble post office closed on December 15, 1917, and Lodi was never the same. An experimental concentrator built there in 1919 was soon abandoned.

The Illinois Nevada Mines Corporation purchased the Illinois mine in 1921. Hughes and Hatterly, two men working for the company, made a rich gold discovery just south of the Illinois mine. The company installed a forty-horsepower hoist, and soon the mine was producing again. Miners worked only the higher levels of the mine, and most of the ore came from the 200- and 300-foot levels. The company also built a water pipeline from Marble Canyon, six miles from Lodi across Lodi Valley. The fresh water flowed by virtue of gravity, and there was a constant supply. From the Illinois mine, the water pipeline is still visible all the way across the valley.

The Illinois Nevada Mines Corporation's period of activity ended in 1928. The last business remaining in the district was the Archibald boardinghouse at Lodi Tanks, which finally closed in 1929. Except for a minor revival during 1940, the district never reopened. Total production value for the district stands at $1.3 million. Burt's faith in the mines remained intact, and he was still in Lodi when he died in 1951. His ashes were scattered at the Illinois mine, which his grandchildren still own.

Remains at Lodi Tanks include a number of brick foundations, the metal water tanks, and corrals. At the Bob site, the ruins are much more extensive. There are two mines, both with complete workings. One building, apparently a blacksmith shop, still stands. A huge pile of wood rubble marks the boardinghouse. It appears to have given out in the late 1970s. An abundance of stone foundations from the earlier operations are scattered in a small arroyo just below the Illinois mine. The ruins of the smelter are slightly to the left of the road heading into Lodi. There are also extensive dumps that make for interesting exploration. Lodi is a fascinating site and should not be missed.

Marysville

DIRECTIONS: *From Ione, continue northeast for 13 miles. Exit right and follow the road for ½ mile. Exit left and follow the road for 1 mile. Exit right and follow the road for 3 miles to Marysville.*

Marysville was a short-lived mining camp that sprang up after prospectors made discoveries in the area in the summer of 1863. By July 1864 miners had laid out a townsite and built three houses, and the camp had a population of twenty. Mines being worked were the Indian, the Empire, the Tampico, the Charleston, and the Shoshone. However, the small deposits that prospectors were finding were quickly mined out, and the mines were all abandoned by the fall. As a result, the budding town also emptied, and the district inspired no further interest. Only the ruins of a couple of stone cabins mark the site, one of the older camps in Nye County.

Midas

DIRECTIONS: *From Idlewild, continue north for 2 miles to Midas.*

The small mining camp of Midas existed from the 1880s until around the turn of the century. It sprang up after nearby Ione became active and some small deposits were located nearby. There was a post office named Midas from 1882 to 1903, but it was located at Ione. There were never more than fifty people at Midas, and only a couple of frame buildings were ever erected there. Midas's population is listed in the 1910 census as twenty-five, but there is doubt as to whether this figure refers to the number of people at Midas or at Ione. It appears that Ione was still considered Midas during the time of the 1910 census.

Limited mining, none of it rich, around the Midas camp could not support the residents, and the site was abandoned with little fanfare. Today only scattered rubble marks the spot. Fresh water is available from nearby Midas Spring.

Milton

DIRECTIONS: *From Ione, head south on Nevada 844 for 6.4 miles. At the point where Nevada 844 heads west, continue straight and follow the road for 3½ miles to Milton.*

Milton was a small vegetable ranch run by a man named Milton. He raised vegetables and sold them in Grantsville and Berlin in the 1880s. However, mining higher up in the canyon first focused attention on Milton in 1863. A small-scale boom ensued, and prospectors built twenty homes near the mines. In the spring of 1865 the twenty-stamp Tanshish mill started on local ore. However, the mill only operated for a couple of months and then closed because the ore was almost valueless. Everyone had left by the fall of 1865. The mill was dismantled, and the houses were moved to Grantsville. Miners worked a small antimony mine in Milton Canyon in the 1930s and 1940s. E. Berryman owned the mine and a number of surrounding claims. Operations were not very successful, and Berryman finally gave up, suffering a significant loss.

At the Milton ranch site, a trailer and one wooden structure still remain.

Nothing is left at the antimony mine. All that marks the old 1860s camp of Milton are the faint foundations of the mill and a couple of stone ruins.

Ophir Canyon

(Twin River) (Toiyabe City) (Ophir City)

DIRECTIONS: *From Round Mountain, head west on Nevada 378 for 2.7 miles. At the junction of Nevada 376, take a right (north). Follow 376 for 16 miles. Exit left and follow the road for 2½ miles to Ophir Canyon.*

A Frenchman named Boulrand first discovered silver in Ophir Canyon in 1863. Joseph Patty, G.H. Willard, and John Murphy, for whom the Murphy mine was named, made additional discoveries in the canyon in 1864. The Twin River Mining Company formed in 1864 and immediately purchased the Murphy mine. The following year, the company constructed an elaborate, $200,000, twenty-stamp mill, which included the first Stetefeldt furnace ever built. The mill also had eight roasting furnaces and eight amalgamating pans.

Demand led to the construction of an $8,000 wagon road from the stage road in the middle of Big Smoky Valley. The road was more expensive to build than it normally would have been because of the nine bridges that had to be erected across creeks that flowed near Ophir Canyon. By 1867 almost 400 people were living in the canyon. Buildings included a church, a number of saloons, a few stores, a school, and fraternal lodges. A post office, with William Smith as postmaster, opened on June 18, 1867, and was called Twin River. The town was known as Ophir Canyon to its residents, who were dismayed because outsiders had labeled it Toiyabe City, which was not the residents' choice. A tri-weekly freight line to Austin was also set up. The cost of shipping was $25 per ton.

The Twin River Mining Company ran into trouble because of the extremely tough rock in which the silver was imbedded. Laborers were sometimes able to extend the Murphy shaft by as little as ten feet after working all day and all night. The company also had to spend large sums each month to sharpen the 18,000 drills used to penetrate the rock. The mine and mill produced $700,000 worth of ore before the company declared bankruptcy in late 1868. Although the mine was very productive, the company was never able to pay any dividends because of the cost of removing the ore. In 1867, for example, the mine produced $427,000 worth of gold and silver and paid extracting costs of $326,000.

When the Twin River Mining Company left in 1868 the town was practically abandoned, although the post office remained open. The Cambridge Silver Mining Company purchased the former Twin River operations in 1869 and began exploring the holdings, but in January 1870 it was forced to shut down because whites—upset over its controversial practice of hiring cheap Chinese laborers at the mines—would not allow its operations to continue. Chinese worked for $1.65 a day, while whites cost $4 a day. The Murphy mine reopened in 1872 after a rich new ore body was discovered in the lowest level of the 500-foot shaft. The *Reese River Reveille* commented on the discovery, reporting that "the Murphy mine will soon rank among the most permanent and best paying mines in eastern Nevada."

The Twin River Mining Company began operating in earnest in 1874, and the mine continued to produce steadily into the 1890s. However, the company decided to sell the mill to Stephen Roberts for $4,500 in May 1875. The Twin River Company was content to lease out its holdings from 1877 until 1885. Leasers' discoveries of additional deposits during the summer of 1885 breathed new life into the company, and during the fall of 1886 it re-timbered the Murphy mine. The company, now renamed the Chicago Mining and Reduction Company, hired eighty men to refurbish and restart the mill. Henry Whitton was president, and D. H. Jackson was superintendent. The following year the company expanded its operations, opening the Grizzly and Cinnamon mines. The company bought the old Prussian mill at Jefferson and installed it in Ophir Canyon. This activity attracted some businesses back to the canyon. Wilson Brougher moved his saloon from Barcelona. Later in

the year he was elected County Sheriff. Thomas Tate's Austin-Belmont stage served the town.

The Chicago company wasn't the only active entity at Ophir. Pat Leonard, James Cruickshanks, H.H. Warne, and John Foster worked a gold mine and built their own three-stamp mill. Signs of trouble appeared in 1887 when the Chicago Company suspended operations in December. While the company resumed work in July 1888, the writing was on the wall. The company folded in late 1888, and the county sold its property in May 1889. A year later the mill restarted to work stockpiled ore, but it only operated for a short while. The town never exceeded a population of 100 after it peaked in 1867. Between 1863 and 1893 prospectors made more than 100 claims and sunk four shafts in the canyon. The value of the district's total production during this period was slightly more than $3 million. By the beginning of 1894 the mines had all closed. The post office shut down on December 5, 1893, signaling the end of Ophir Canyon.

The canyon remained quiet until 1917, when the Nevada Ophir Mining Company purchased eight claims, including the Murphy mine in Ophir Canyon. The company was incorporated in Utah in 1917, with Walter Trent as president. Claims included shaft mines of 500 feet (the Murphy mine) and 300 feet. The Murphy mine had branching tunnels every 100 feet, including one on the sixth level that was well over 700 feet long. In 1918 the Nevada Ophir Company built a small stamp mill, which operated until 1923. The company suspended all operations in Ophir Canyon in 1923 and then reopened the next year. It built a small cyanide mill, but the venture proved unprofitable. The company folded in 1925.

Since 1925 Ophir Canyon has seen only very limited activity. What little has occurred has been conducted by small, independent groups. After the Nevada Ophir company folded, only one other mining company ventured into the district. The Ophir Canyon Mining Company formed in 1935 and tried to make a profit for the next two years by working the Murphy mine. Walter Trent, who had owned the mine since 1916, was president of the company. After sustaining losses for two years, the company folded in 1937. In 1951 and 1952 the Newmont Mining Company explored the area looking for tungsten and began developing the Bobby Tungsten mine, but the venture proved unsuccessful. During the late 1970s and early 1980s miners worked a small mine near the entrance of the canyon, but the returns were insignificant.

The best remains in Ophir Canyon were the Murphy mine and mill, located up in the canyon. Unfortunately, the forest service deemed the beautiful stone buildings a danger to public safety and razed them. Other remains include ruins of the stamp mill that was constructed later, some structures, and the workings of a few other mines. A number of stone ruins sit on the north side of the canyon. A small, recently rehabilitated graveyard at the mouth of Ophir Canyon contains many unmarked graves along with a few forlorn, marked ones. The remains are fairly rewarding, but a visit to the town is less interesting than its history is.

Orizaba

DIRECTIONS: *From Black Springs, head south on old Nevada 89 for 4½ miles. Exit right and follow the road for 1 mile to Orizaba.*

Ed Workman, who later discovered National, came across gold and silver ore in the western part of the Cloverdale district in 1909. He was soon shipping ore from the Orizaba mine to the town of Millers, in Esmerelda County. The property came under the control of the Diamondfield Black Butte Reorganized Mining Company, based in Goldfield. The company was incorporated in 1905 and then reincorporated in 1910 with $2 million in capital. The Diamondfield Company sold its claims to the Orizaba Mining and Development Company but maintained a financial interest in the new company, which was incorporated in 1915 with $1 million in capital.

In 1911 prospectors in the area staked six claims covering 120 acres. The ore contained quantities of gold, silver, lead, and copper. By October 1915 miners had uncovered over 1,500 tons of rich ore in the Orizaba mine and processed another 1,200 tons of lower-grade ore in the tailing piles. During the next two years the Mines Selections Company gained control over all the stock and assets of the Orizaba company and became the property's new owners. In 1917 the company moved mining equipment from Golden to improve the Orizaba mine. Orizaba's main disadvantage was the fact that 100,000 gallons of water were seeping into the mine every day.

By 1918 the holdings of the Mines Selection Company had expanded

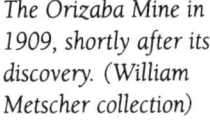

The Orizaba Mine in 1909, shortly after its discovery. (William Metscher collection)

to nine claims covering nearly 180 acres. It is interesting to note that the Diamondfield Company owned most of the Mines Selection Company's stock. The district production level continued to increase for a while, but by late 1918 only an estimated $18,000 worth of ore remained in the mine veins.

The company brought modern equipment in by freight wagons from Millers, in Esmerelda County, but the ore was rapidly disappearing. Six months later the area was almost completely deserted. From 1913 to 1918 the mine produced ore valued at $122,000. In May 1922 the Silver Mines Selection Company, which was unrelated to the Mines Selection Company, purchased all holdings at Orizaba and began removing water from the mine. John Highland, who was involved in mining at nearby Republic, discovered the Gold Exchange mine in October. The Tonopah Mines Syndicate bought out the Silver Mines Company in April 1924 but met with little success and left in 1926. Hubert Welch last worked the mine in 1949, but he gave up in less than a year. The value of the district's total production was close to $128,000. This amount might seem quite small, but silver, the mainstay of the Orizaba ore, was only worth sixty cents an ounce at the peak of the mine's activity.

Fairly extensive ruins still remain at Orizaba. Alongside the remains of the two larger mines, three structures struggle to stand. A quarry is located a quarter of a mile past Orizaba. In 1980 Allen Coombs and Jim Larson built a small mill there, and it has been used sporadically since.

Pactolus

DIRECTIONS: *From Gabbs, head south on Nevada 361 for 5 miles. Exit left and follow old Nevada 89 for 16.4 miles. Exit left again and follow this road for 2¾ miles to Pactolus.*

Pactolus was an extremely insignificant and short-lived gold-mining camp. L. K. Mau, Curley Jones, Dave Long, and Jack Lee initially discovered high-grade gold in July 1902, prompting a camp of twenty to form. On August 27, 1904, the government awarded Pactolus a post office, and A. H. Mann was postmaster. But the office closed on December 13 because interest in the camp had faded. Sam Eva of San Francisco owned most of the patented claims around Pactolus. The Pactolus Mining Company owned the only major mine—the Doctor mine—which yielded several shipments of $75-per-ton ore from 1905 to 1907. The shipments became smaller and smaller until the mine finally closed.

The camp quickly emptied after the Doctor mine ceased to operate. In July 1908 the Gold Crown Mining Company agreed to buy the mine for $125,000, but it withdrew its offer when the owners raised the price to

$175,000. As a result of this burst of interest, the Sierra Vista Mines Company formed and began work in the Everett and Butler mines, which employed eighteen.

Nothing more of note occurred at Pactolus until November 1919, when Senator L.A. Friedman, J.T. Foodin, Bert Downing, and Fred Hill bought the Pactolus mines. In the 1920s new discoveries lured more than thirty men back to the camp. Miners constructed a number of new buildings, and the old ones were used for firewood. Activity was extremely limited, however, and no production was recorded. No other mining activity has taken place in the area since. Only one dilapidated building marks the site.

Paradise Peak
(Jim Graham's Camp)

DIRECTIONS: *From Goldyke, head directly east on a good road. Follow it for 7½ miles to Paradise Peak (elevation 8,650 feet).*

Paradise Peak, the largest mountain in the area, was the site of Jim Graham's Camp, named after the discoverer, which first formed in the teens. Graham had been prospecting in the region since 1863 and made his finds on Paradise Peak in 1864. He had earlier discovered ore in the Mammoth District. A.T. Hatch and James Kirkpatrick formally organized the Paradise Mining District on April 6, 1865. Hatch was elected district recorder. Within a couple of years, however, only Graham remained in his camp. Along with his partner, A.B. Millett of Smoky Valley, Graham worked his mine until 1888. During the 1890s prospectors dug the North Star mine 1,000 feet into the side of the mountain, and the mine yielded low-grade ore that had low values in gold, silver, and copper. W.C. Peay, who owned the Rocket mine, made some new discoveries in the area, which led to the formation of the Rich Hills mining district in March 1910. Will Hanby was the recorder. But the mine never produced and was soon abandoned. The area remained quiet until 1926, when the Paradise Peak Mining Company began to rework the North Star mine. John Reinmiller of Reno organized the company, which was incorporated that year. The company mined ore that assayed from $2 to $18 per ton. The company continued to operate for a few more years, but the low-grade ore was not sufficiently valuable to support its continued operation.

Prospectors discovered mercury at the site in the mid-1930s. The North Star mine was renamed the Scheebar Mercury mine, and full-fledged operations soon began. The year 1936 was a big one for mercury mining, and more than fifty flasks were produced. In 1943 the Scheebar Syndicate, headed by Benjamin Parker and Julius Redelius, formed and took over operations. The syndicate mined more than 100 tons of cinnabar ore, averaging about sixteen pounds of crude mercury per ton. The company remained active in the district until the late 1940s. In 1980 collapsing remnants of two tunnel mines and the rotting ruins of a number of small wood buildings still existed at the site.

In 1985 FMC began developing a huge mining operation at Paradise Peak to produce gold, silver, and mercury, spending $100 million on the start-up project. Production began in 1986, and the mine was yielding 4,000 tons of ore a day. The mine and mill produced 986,000 ounces of gold and a little more than 19 million ounces of silver between 1986 and 1990. Most of the company's large work force resided in nearby Gabbs. The enormous operation overtook the remains of Jim Graham's Camp. The mine was mined out in 1993, and the mill closed in May. Between 1986 and 1993 the district produced 1.58 million ounces of gold and 24 million ounces of silver. The huge open pits remain today. The site is fenced and off-limits.

Park Canyon

DIRECTIONS: *From Round Mountain, head west on Nevada 378 for 2.7 miles. Exit right (north) on Nevada 376 and follow it for 21.4 miles. Exit left and follow the road for 1 mile. Take the right fork and follow it for ¾ mile to Park Canyon.*

Park Canyon, a small silver camp, originally formed in 1865, soon after Broadhead and Davis discovered the Buckeye mine. While the Buckeye was in Summit Canyon, which is south of Park Canyon, Park Canyon's superior location, along with other discoveries made in the area, put it at the center of the North Twin River Mining District, which had formed in 1863. By August 1865 mines in Park Canyon included the Twin River Giant, the Giant, the Fairmont, and the Bigler.

The Buckeye Mining Company of New York ran the Buckeye mine, whose ore was primarily silver but also contained values in gold, lead, and copper. Other mines in Summit Canyon included the St. Paul, the Lexington, the Troy, the Wide West, the Santa Rosa, the Summit, the St. Louis, the Sultan, the Albany, the Monarch, the Antelope, the Crown Point, the Gettysburg, the Edinburgh, the Tecumseh, and the Brooklyn. The silver ore occurred in pockets, however, and its value was extremely inconsistent. The ore was shipped to Austin until 1867, when the La Plata Company moved a ten-stamp mill from Yankee Blade (in Lander County) to Park Canyon. The La Plata Mining Company of Reading, Pennsylvania, housed the stamps in a building of native brick and stone. The mill opened in July 1867 and operated regularly until it closed two years later.

The camp that formed around the Buckeye mine contained a number of stone buildings and little else. It has long been rumored that in 1867 a deposed Hawaiian queen, Queen Liliuokalani, lived in Park Canyon and owned a few mining properties. After the La Plata mill closed, Park Canyon emptied.

The district experienced a number of revivals after 1869, the most significant of which occurred in 1885 and 1886. The Buckeye mine reopened, and miners worked a few small surface operations. In July 1885 the Giant Silver Mining and Bullion Company, with W. E. Darwin as president and John Truman as superintendent, formed to work the Giant mine. The old Jefferson mill moved to Park Canyon, but the ore was impossible to work properly. In an attempt to solve the problem, the company bought the furnace of the Citizen's mill in Austin and added it to the mill in Park Canyon. It too was a failure, however, and the company curtailed operations in late July. The last business in Park Canyon—the Truman Restaurant—finally closed in 1890. The mill and furnace sold for $500 in 1891. A post office, with Stephen Truman as postmaster, opened on January 25, 1886, but closed on November 11.

Park Canyon remained a lonely ghost town until 1905, when it once again began to show signs of life. This revival turned out to be its longest period of sustained activity. It lasted until 1916, and before it faded miners had removed a sizeable amount of ore from the area's mines. The revival centered on the activity of the Albion Millett Gold Mining Company, of which George Bartlett was president, on the Brooklyn group. In December 1911 the company sold the Brooklyn mine to the Lucky Dick Gold Mining Company. Soon afterwards the Smoky Valley Gold Mining and Milling Company bought the rest of Millett's claims for $20,000. A new five-stamp mill, built by the Nevada National Company, started in February 1912 and processed ore until 1913. The camp grew a bit, but most of the miners resided at Millett, where a small town had formed. The camp in Park Canyon was sometimes referred to as Millett, which created a certain amount of confusion. D. D. Sullivan and J. W. Lonn restarted the old mill that had been built in 1885, but once again it proved a failure at working the Park Canyon ore. After 1916 Park Canyon faded into the past. The Round Mountain Mining Company bought the mill in 1920 and moved it to Round Mountain.

The only additional activity to take place at Park Canyon occurred from 1937 to 1941 on the Giant Claim, one and a half miles west of the old Park Canyon mill site. The mine was far up on the side of a steep mountain. An 800-foot tram, equipped with a ten-ton ore bin, was rigged to bring the ore down from the mine. The mine produced $12,000 worth of ore before closing in 1941. The road to the Giant Claim site is washed out, and a three-quarter-mile hike is required to reach the spot. Remains of the tram line and the mine are still visible. During the early 1950s Louis and Phil Meyers worked some

Preserving the Glory Days

The only building left in Park Canyon. (Shawn Hall collection, Nevada State Museum)

gold claims in the area, and Warfield, Inc., explored it looking for tungsten, but no production took place.

Ruins in Park Canyon are scant. The Nevada National mill, near the mouth of the canyon, is in fairly good condition, but the stamps have been removed. There is only one standing structure—a small stone cabin. The La Plata mill ruins consist of stone foundations. The only other signs of past activity are a few scattered stone ruins and piles of old tin cans.

Peavine

DIRECTIONS: From Manhattan, head west on Nevada 377 for 6.8 miles. Exit left (south) on Nevada 376 for 2.8 miles. Exit right and follow the road for 4 miles to Peavine.

Peavine is a ranching settlement that dates back to the 1870s, when Lemuel (Lem) and Johanna Compton settled there. For many years it was a stop on the Tate and Wallace Belmont–Sodaville stage line. By 1881 the small settlement had a population of twenty-five and was home to miners and mill workers from the surrounding area. A post office, with John Campton as postmaster, opened on August 25, 1890, and remained in operation until October 5, 1890.

In 1893 Lem suffered a severe stroke, which left him seriously incapacitated. Johanna ran the ranch and stage station by herself. Beginning in 1867,

before she and her husband had bought the Peavine Ranch, Johanna, who was then married to Frank Bradley, had run a small hospital at Belmont. She had previously served as a Union nurse during the Civil War. After visiting Peavine in 1899, Tasker Oddie made the following observation about Johanna: "She is the greatest character in Nevada, is 55 years old, weighs 225 pounds, talks 3,000 words a minute, too, never stops, scolds the old man continuously, is the most hospitable creature in the world, wears the largest boots of any man or woman in Nevada, swears like the Devil, is always gossiping, and works in the fields like a man." The letter in which this statement appears can be found in Oddie's *Letters from the Nevada Frontier, Correspondence of Tasker L. Oddie 1898–1902.* Oddie was a frequent visitor of the Comptons and always looked forward to seeing them.

A dreadful fire hit the Compton family hard in November 1900, destroying everything but the house. Lemuel was not at home when it struck, having been taken to Belmont to vote. Johanna apparently tried to save the buildings and was later found burned to death in the rubble of the stable. Johanna, who served as the local doctor, was known far and wide as a genial host, and everyone mourned her passing.

For many years the Peavine Ranch had a small supply store that served local prospectors and other ranches along Peavine Creek. The ranch has maintained a low-level existence, with its population never exceeding twenty-five. It is still in operation today. A number of old stone buildings are nearby.

Preserving the Glory Days

A campground is located a few miles past the ranch. If you choose to camp there, bring your own firewood, for there isn't any available.

Penelas

DIRECTIONS: From Ione, head northwest on a good road and follow it for 9½ miles. Exit left and follow the road for 2 miles to Penelas.

The Penelas mine, which Severino Penelas originally discovered in the teens, was the biggest producer of the Bruner Mining District. Penelas first came to Phonolite in 1913, and for almost twenty years he lived in abject poverty, working in the mine by candlelight and without ventilation. He died of pneumonia that he contracted while in the mine. After his death it was discovered that he had hoarded almost $100,000, all money earned from his mine.

Other than the Penelas mine, no activity took place in Penelas before 1930. Major operations began there in 1931. The Penelas Mining Company, headed by L. D. Gordon of Reno, purchased the mine in 1931 from the Penelas estate and continued to work it until 1942. In January 1936 the company completed a fifty-ton cyanide mill that gained a reputation as one of the most efficient mills in the state. By the end of 1936 there were thirty buildings at Penelas. By 1939 the camp included twelve homes and six bunkhouses, which housed fifty employees and was called the most modern mining camp in Nevada. At the peak of the mine and mill's operation, the camp had a population of seventy. Most of the residents left when the mill closed in 1940 after the Penelas mine's ore values drastically declined. When the mine closed in 1942 the camp was quickly abandoned. At that time the Penelas mine was over 900 feet deep, had almost 5,000 feet of lateral work, and had produced close to $900,000 worth of gold and silver. The mill was dismantled and removed in late 1942. All the other buildings were moved to other places, including Gabbs, during the ensuing years. Nothing at all remains of the camp. Only mill and building foundations and mine tailings mark the site.

Phonolite

(Bruner)

DIRECTIONS: *From Gabbs, head north on Nevada 361 for 3 miles. Take a right onto Lodi Valley Road and follow it for 10½ miles. Exit right and follow the road for 2½ miles to Phonolite.*

A group of prospectors who had been working the northern Pacific range discovered gold and silver ore at Phonolite in July 1906. Ore from the Paymaster mine assayed at more than $2,500 per ton, setting off a small rush to the area.

Two major forces set up shop at the Phonolite townsite. Henry Bruner, originally from Kansas City, was active in promoting the townsite and was named postmaster when the Phonolite post office opened on January 26, 1907. The Phonolite Townsite, Water & Light Company was also a heavy promoter of the site. The company had been incorporated in 1907 with $1 million in capital. Promotion of the townsite included plans for a city water system and an electric plant. The streets were named for states and the cross streets for minerals. The plans fell through when the promoters left, however. By 1909 the district had lived up to its name: Phonolite had turned out to be

Mining operations of the Kansas City–Nevada Consolidated Mines Company in 1917. (Firmin Bruner collection, Central Nevada Historical Society)

Preserving the Glory Days

a phony—all talk and no ore. The post office closed on July 23, 1909. Only Bruner and Larry Ryan, an old-timer who had claims in the area, were left.

It was not until late 1910 that Phonolite again saw some activity. Bruner discovered new deposits, and by 1911 more than twenty people were in the district. The post office reopened on October 17, 1910, and was renamed Bruner. Bruner was postmaster once more. Activity was limited, however, and the office closed a second time on January 31, 1912. Two new mining companies, both of which Bruner ran, formed the following year. The two companies, the Phonolite Paymaster Mining Company and the Phonolite Silent Friend Mining Company, carried on operations until 1915, when they merged to form the Kansas City–Nevada Consolidated Mines Company, which was incorporated early the next year with $6 million in backing capital. The president of the company was Walter Neff, and Bruner served as secretary. This new burst of activity justified the reopening of the Bruner post office on December 28, 1915, and the office continued to operate until June 15, 1920. Bruner was postmaster yet again.

The Kansas City–Nevada company owned twenty-five claims in the district. The biggest producer was the Paymaster mine, which was 300 feet deep and had more than half a mile of lateral work. The ore was rich enough that the company built a fifty-ton mill in June 1919 to treat it. The mill was located a short distance away from the original Phonolite townsite, which essentially moved to a new area adjacent to the mill. The company invested almost $300,000 in the mine and mill. The money was well spent, and the mine paid back its investors. In 1921 the company completely remodeled the mill and made it much more efficient. In addition to treating the Bruner district ore, the mill also had a contract with the Broken Hills Silver Corporation, located twelve miles away.

The company ran into financial trouble in 1924 and was forced to fold. Walter Neff maintained control over the mineral rights, and in 1925 he organized the Golden Eagle Mining and Milling Company. That company worked the old Kansas City–Nevada Company property. A $90,000, four-foot water pipeline was constructed eight miles across Lodi Valley to Lebeau Creek, but the water supply ceased to exist during the summer months. The Golden Eagle mill began operating in 1926. During this time, the camp consisted of a boardinghouse and seven cabins. The company worked the Golden Eagle property at a low level through 1929 before it gave up and left the district. The fifty-ton mill burned down in early 1930. With the mill gone, Phonolite inspired little further interest.

There are few remains at Phonolite. The huge concrete foundations of the mill dominate the site. The ruins of two wood cabins are nearby. The Paymaster mine is only 200 feet from the mill ruins. Since the 1980s Jessie Wilson has been carrying on a low-level operation, running the Paymaster mine and a small mill. The site is marked No Trespassing, so be sure to ask at the mine for permission before exploring. In 1990 Newmont Exploration and the Miramar Mining Company outlined a microscopic gold deposit named the Bruner Project. It is possible that at some time in the future the Bruner district could produce riches once again.

Potomac
(Cimmeron) (San Lorenzo)

DIRECTIONS: *From Liberty, continue north for 3½ miles. Exit right and follow the road for 4 miles to Potomac.*

Potomac, one of the first camps in Nye County, formed in October 1863. The discoverer was J. P. Cortez, who had also discovered La Libertad, also known as the Liberty. Cortez named the small camp San Lorenzo, after the nearby spring. Two arrastras were constructed at the spring to process the ore from the San Lorenzo mines and the nearby Liberty mine.

Cortez's activity ended in the 1870s, and the small camp saw only intermittent activity afterward. The site became fairly active from the 1930s to the early 1950s when the owners, E. M. and Margaret Booth, employed a small crew to work three mines and leased out other properties. In July 1936 the Tonopah Mining Company bonded six claims from the Booths. At the same

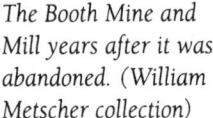

The Booth Mine and Mill years after it was abandoned. (William Metscher collection)

time, Jake Stephenson and Jack Gray were working their own claims. In June 1938 the Booths sold a block of their claims to Charles Taylor, who organized the Pacific Buttes Mines Company. He sold a portion of the claims to the Silver Divide Mines Company the following month. But within two years the property had been abandoned and had reverted back to the Booths. A number of wood cabins were built at Potomac during this period, and all of them are still standing. Work on the shafts totaled more than 700 feet, and the mines produced more than $15,000 worth of ore before they closed.

Remains at Potomac are fairly extensive, although they are quite modern. In addition to the cabins, there are also mine ruins at the site. The Potomac mill, built by the Booths, still has its galvanized steel covering. Inside are the remains of a small, crude concentration system. Be careful when driving to Potomac; the road is extremely rough and sandy.

Pueblo
(Gold Center)

DIRECTIONS: *From Round Mountain, head west on Nevada 378 for 2.7 miles. Exit right (north) on Nevada 376 and follow it for 21.4 miles. Exit left and follow the road for 0.1 mile. Take the left fork and follow the road for 1½ miles to Pueblo.*

Pueblo formed in 1863 shortly after discoveries were made in nearby Park Canyon. Nothing substantial was discovered at the camp, and the site was soon abandoned. It was not until 1905, when a wandering prospector found a skeleton of a man next to one of the two streams near the old Pueblo townsite, that the town came back to life. An old gold pan was lying next to the dead man. The prospector panned the dirt, and it assayed at $5,000 a ton. Within a matter of weeks a camp of sixty tents had sprung up. Miners laid out a townsite in March 1906. A rival townsite, named Gold Center, was platted nearby, but it faded even more quickly than Pueblo did. By 1906 more than a dozen wood-frame buildings had been built, and the town's population had risen to almost 500. The Marvel Mining Company and the Pueblo Gold Mines Company were both incorporated and began working a number of claims in the area. Unfortunately for Pueblo, however, the veins discovered in 1905 were very shallow and had been worked out by the next summer. The town emptied as fast as it had filled, and within weeks the site was completely abandoned.

Not much of Pueblo remains. A few dilapidated wood buildings still struggle to stand, and a couple of old wrecked cars are scattered among the ruins. Fresh water is available here.

Quartz Mountain

DIRECTIONS: *From Gabbs, head north on Nevada 361 for 9 miles. Exit right and follow the road for 3½ miles to Quartz Mountain.*

Initial discoveries in the Quartz Mountain area took place in early 1920, but no activity occurred there until 1925, when work began on three mines, one of which contained a high-grade ore shoot of lead and silver. This started a small rush to the district. Within months almost 500 people were gathered around the base of Quartz Mountain, and by June 1926 E. S. Giles, president of the Quartz Mountain Townsite Company, had platted a townsite. An airplane-landing field was graded, and 440 lots went up for sale on June 27, which the company celebrated as Townsite Day. The company moved six buildings to Quartz Mountain from Rawhide during Quartz Mountain's first week of existence. The town's first business was the Dodge Brothers' mercantile store. Soon the booming town had its own newspaper, the *Quartz Mountain Miner.* Miners moved a number of buildings to Quartz Mountain from other towns that had died. The town of Rawhide, deserted after the disastrous fire of 1908, supplied the town with a few buildings, including a saloon and a barbershop.

Automobiles were plentiful in Quartz Mountain, making it relatively easy for anxious prospectors to travel to the site. Miners used trucks to haul out the mined ore and to bring back supplies and water, the latter of which was in short supply at Quartz Mountain. The mail came via Broken Hills (in Mineral

County), which the Quartz Mountain boom had revived. By the end of the summer of 1926 the town had a number of cafes and saloons, three grocery stores, two barbershops, and five general merchandise stores. In response to pressure from residents, a post office, with Everett Maupin as postmaster, opened on June 7, 1927.

More than fifteen mining companies were active in the district between 1926 and 1930. Some only maintained a claim or two, while others were large operators. In 1926 four mining companies worked the area, but all of them folded within a year. One of them, the Goldfield Quartz Mountain Mining Company, built a small mill in May 1926 to treat ore from the company's two mines. The company gave up in the fall of 1926. George Wingfield, a prominent Nevada mining businessman, owned the San Felipe Mining Company, which controlled fourteen claims in the district. In September 1927 the property was transferred to the San Rafael Consolidated Mines Company, which was incorporated in August 1927 as the consolidation of four mining companies that had previously worked the district. The four companies were the San Rafael Development Company, the Calico Quartz Mountain Mining Company, the San Felipe Mining Company, and the Exchequer Quartz Mountain Mining Company. The property of the San Rafael Consolidated Mines Company encompassed twenty claims in the district. There were two shafts among the claims, one 600 feet and the other 400 feet, and both combined had 6,500 feet of lateral workings.

Another major company working the district was the Quartz Mountain Mines and Milling Company. Incorporated in January 1928, it gained control over the Quartz Mountain Metals Company and the San Rafael Consolidated

The collapsed remains of a hotel at Quartz Mountain. (Shawn Hall collection, Nevada State Museum)

Only building standing at the Quartz Mountain townsite. (Shawn Hall collection, Nevada State Museum)

Overview of Quartz Mountain in 1925. (Central Nevada Historical Society)

Mines Company and discovered a new vein, the Lease. This vein, along with the other claims, produced more than $300,000 worth of ore. The company operated three major mines. The San Rafael mine, which it dug to a depth of 450 feet following the discovery of the Lease vein, produced more than $250,000 worth of gold and silver from 1920 to 1927. The Calico mine was a little more than 400 feet deep. The third mine, the Quartz Mountain Metals mine, was just under 300 feet deep. Although the properties looked promising, the ore deposits were so small that by 1930 the company had left the district.

A few other small companies also worked Quartz Mountain in 1929, but life seemed to be ebbing from the town. The post office closed on January 15, 1929, and soon Quartz Mountain was empty. Its ghosts have rarely been disturbed since. Until recently a number of buildings still stood at the site, but only two wood cabins remain today. The townsite is very easy to locate, for it is marked by large piles of wood rubble from the recently collapsed buildings, one of which was once a two-story hotel. At least ten collapsed buildings are distinguishable on the flat below the Quartz Mountain mine. Ruins from the mines abound in the area and are well worth investigating. Plan to spend a long time at Quartz Mountain, for the site is spread out, and none of it should be missed. Even though it lacks many standing buildings, Quartz Mountain is one of Nye County's most interesting ghost towns. The road to the site is quite sandy, so exercise caution.

Ray

DIRECTIONS: *From Tonopah, take US 95 west for 1½ miles. Exit right and follow the road for 4 miles. Exit left and follow the road for 5 miles to Ray.*

Judge L. O. Ray, who found rich silver ore in the mountains north of Tonopah, discovered Ray on Christmas Day 1901. He filed for six claims in the area, naming them the Hornsilver, the Ladysmith, the Boomerang, the Jenny Belle, the Christmas Gift, and the Teddy Roosevelt.

Judge Ray was originally from Iowa. After graduating from college, he worked in Colorado as a mining engineer. In 1901 he arrived in Tonopah, where he met Jim Butler. The two men worked together for a few months before the judge struck out on his own. Before he had made the discoveries at Ray, the judge had already amassed a small fortune selling rich claims that he had filed on around Tonopah.

In February 1902 miners discovered ore assaying at $240 a ton at Ray. Soon almost 200 miners flocked to the area to work the new mines. Judge Ray decided to sell his claims in April 1902. The Ray and O'Brien Gold Mining Company purchased the holdings. Because of the judge's faith in his new camp, he accepted Ray mining stock as the bulk of his payment.

The Ray and O'Brien company had a very strong board of directors, including Jim Butler as president and Judge Ray as general manager. The company immediately began intensive prospecting in the Ray area and soon uncovered a rich new ore body. This led to the most extensive development of the Ray

Main camp of Ray in 1901. (Nevada Historical Society)

area. The largest mine at Ray had a 300-foot shaft, which had a 100-foot air shaft. At the 160-foot level of the shaft, stoping of ore deposits was instituted. The ore was rich not only in silver but also in lead.

At this time the Ray Mining Company, with Jim Butler as president, formed to work the Hornsilver mine. In June the Fraction Mining Company, later renamed the Coronation Mining Company, purchased the Magpie, the Nancy Hanks, the Rex, the Laura B., and the Christmas Gift mines. Tasker Oddie, Harry Ramsey, and L. O. Ray owned the company. The Ray Extension Mining Company, of which Clay Peters was president, began working the Lone Jack, the Ibex, and the War Path mines.

Because of Ray's location, it was difficult to transport silver ore to Tonopah. Wagons were the only available mode of transportation, and the condition of the road leading from Ray to Tonopah made even this type of travel almost impossible. Plans were made to build a suitable road to Ray, but before they could be carried out the ore at Ray faded, and they were scrapped. Ray delivered its first shipments in February 1903 after the ore began fading, and they were the last gasp of a dying town. But some people, hoping that interest in the camp would revive, stayed at Ray, and a large Christmas celebration was held in 1903 at the town's only frame building, the Kunize brothers' so-called "mansion." In June 1904 the Ray and O'Brien Mining Company and the

Ray-Tonopah Mining Company consolidated to form the Ray Consolidated Gold Mining Company. The following month the Mogul Mining Company began work on the New Life, the Reliance, the Damon, the WJP, and the JMP mines. The Ray Extension Company was renamed the Toledo-Tonopah Mining Company, but it folded soon afterwards. Despite all this activity, once again Ray's mines produced little, and by the end of 1904 only a handful of prospectors were left. It wasn't until 1915, when Julius Goldsmith began working the old Lee Bell property, that any additional work took place. In 1921 Walter Ross and Arthur Borien relocated the old Mogul property but found little paying ore.

Because Ray's life was so short, no substantial buildings were erected in the town. Absolutely nothing remains there except for faint depressions at the site. Small tailing piles on the hillside mark Judge Ray's claims. The road to Ray is extremely rough, with many sandy areas.

Republic

DIRECTIONS: *From Black Springs, head south on old Nevada 89 for 3.1 miles. Exit left and follow the road for 2 miles to Republic.*

Republic was one of a number of very small camps that sprang up in Nye County wherever a mine was located. The Republic group consisted of four mines. The Cirac mine, the richest and oldest of the four, was a fair producer in the early 1900s but panned out in little more than two years. L.V. Cirac had discovered the mine in late 1905. Cirac organized the Cirac Mining Company, of which M.S. Bonnifield was president, to work the Hyland, the Workman, the Talty, the Cushing, and the King Baldwin claims. In March 1906 C.A. Norcross and H.B. Tonkin bought five of Cirac's claims for $15,000, but the mines produced little before the district was abandoned in 1908.

In March 1913 miners made new strikes in the Cirac, and Patsy Clark immediately secured a lease. A rush developed as word got out about the $800-a-ton ore. On March 13 the Republic Mining District formed. S.V. Cirac was the recorder. C.S. Wilkes surveyed a townsite, and the town's first business was a saloon run by Newt Crumley. By May there were forty tents and frame houses at Republic. The town received a big boost when Thomas Murphy, widely known as the "father of Goldfield," bonded the Hyland mine, but the excitement died down quickly when the shallow ore pockets ran out. By 1916 only the Rosalie Mining Company was active in the area. It wasn't until 1928 that new finds breathed some life back into Republic. The Ciracs resumed work on the Cirac mine and in December completed a four-stamp, twenty-ton mill. Tom Hyland also began working the area again. He sank a 200-foot shaft and did a small amount of lateral work. The ore from the mine had from eight to twenty-two ounces of silver per ton. William Farris sank a shaft half a mile west of Hyland's mine and mined twelve tons of ore, which averaged four ounces of silver per ton. The activity attracted outside interest, and by 1934 the Triangle Mines Company, the American Gold Mines Corporation, the Tonopah Republic Mining Company, the Wilson Divide Mining Company, the Broken Hills Silver Corporation, and the North Divide Extension Mining Company were all working leased claims. But the flurry of activity produced little, and by 1937 all the claims had reverted back to Hyland and Farris. After they left the district in the late 1940s only limited activity took place. In addition to the mine ruins, a couple of buildings still stand along the canyon road.

Royston

(Quincy) (Hudson)

DIRECTIONS: *From Tonopah, take US 95 west for 2.4 miles. Then take old Nevada 89 north for 23.2 miles. Exit and follow the road for 5 miles. Exit right and follow the road for 2 miles to Royston.*

Royston was a small boomtown that quickly formed in 1921 when word spread that $20,000 worth of silver had been taken from a twenty-four-foot shaft in only six weeks. The camp was originally called Quincy, or Hudson, but its name changed to Royston in honor of W. H. Royston, manager of the Hudson company, soon after the discovery was made. The Hudson Mining and Milling Company staked seventeen claims, including the Golden Eagle and the Best Chance mines, in the district. The Walker brothers of Salt Lake City incorporated the company in April 1921. The company had actually been organized in 1919 but at the time was not large enough for incorporation. Another pre-boom mining company was the Quincy, which operated intermittently from 1904 to 1912 before folding. The property was sold against a judgment of $19,000. In 1913 all mining claims in the district were deeded to McCormick and Company. In 1921 J. R. Walker bought the claims, which Royston had relocated in the meantime. To solve legal problems, the Hudson Company was formed on a fifty-fifty basis between the two groups. The Hudson Company had leased a few of its claims to the Hudson Leasing Company in 1919. That company ran at a loss and folded the same year. The Hudson Company then began its own serious operations. It discovered a vein assaying at $90 a ton and before the boom of 1921 had removed more than $40,000 worth of silver.

By the end of 1921 Royston had a population of 300 and a small business district. Its first frame building housed the Draper Restaurant, and by the end of November sixteen buildings had been moved to Royston. The *Manhattan Magnet* declared that "the new camp of Royston grows like a rank weed." A planned post office did not receive government approval before the town was practically abandoned, and it never materialized. In December 1921 the Royston Piedmont Mining Company formed after purchasing the claims of Briz Putnam and W. C. McGregor. The Super Six Mining Company began ore shipments in January 1922, and the Ben Hur Mining Company also became active.

Some of the mines, including the Golden Eagle, remained open and operated periodically. The Super Six Mining Company leased a number of claims from the Hudson Company but gave up in 1925 after a bout of unprofitable activity. Subsequently miners only worked the Golden Eagle mine. B. F. Betts leased the property after the Super Six Company folded. He worked the mine on a small scale until the Western Leadfield Mining Company purchased the property from the Hudson Company in 1926. The new company enlarged the Golden Eagle shaft to 300 feet and did more than 1,200 feet of branch work. Betts continued prospecting in the area and in 1927 located another rich claim just south of his old one. He leased it to Frank Tabakacci. The Western Leadfield Company gained control of the property in late 1927 and sank a 350-foot shaft, but it abandoned the project as the ore values decreased.

The last company to work the district was the Royston Turquoise Mines Company. The company, which Bryce Sewell and Frank Keller organized in 1929, obtained a permanent lease on some holdings of the Hudson Company, which was not producing but which was nevertheless conducting extensive development. Heads of that company still maintained a one-fourth interest in the Royston Turquoise Company. In 1930 the company merged with the Royston Royal Blue Turquoise Mines Company. Both companies folded in 1931, however, and the district was quiet until 1948, when the Royston Coalition Mines Company engaged in some small-scale activity. After work ceased in late 1948 the district was abandoned forever. It had produced a total of $160,000 worth of ore.

Even though mining activity took place in the Royston district after 1923, the townsite was never again heavily populated. The remains of Royston are almost nonexistent and are extremely hard to locate. The only site markers are tailing piles from the various mining ventures. On the flat below the remains of the Golden Eagle mine are scant wood scraps—the only remnants of the town of Royston. The site is not very interesting and is not really worth the long trek that must be made to get there.

San Antonio

(San Antonia) (San Antone)

DIRECTIONS: From Tonopah, head east on US 6 for 5½ miles. Then take Nevada 376 north for 33.7 miles. Exit left and follow the road for 3 miles. Exit left and follow this road for 7 miles to San Antonio.

San Antonio was a small stage station built around a group of springs. As early as 1845 Frémont had stopped at the springs on his way to meet Kit Carson. San Antonio was a major stop on the Westgate-Ione-Belmont stage line. After gold was discovered in the nearby San Antonio Mountains, the stage stop became a small town as miners began to move to the district. John Courter, who also owned the Potomac mine, built a two-story station/hotel in San Antonio in 1865. He added a forty-foot brick addition in 1879. The district's main mine was the Liberty. A stamp mill, built at San Antonio to treat some of the mine's ore, operated in 1866 and was then moved to Belmont, where new strikes had been made. The Pioneer mill, completed in June at a cost of $120,000, weakened after treating only 100 tons of ore. The mill owners spent another $20,000 before abandoning it in September.

San Antonio soon had a fairly large business district and a population of about 100. A freight line to Austin began running, and shipping costs were $30 a ton. Even after the Liberty mine closed in 1868 San Antonio remained active. The population dropped to less than fifty, but the stage station was still an important stop for travelers heading toward new strikes in southern Nevada. The San Antonio post office, with John Mitchell as postmaster, opened on May 14, 1873, and served the district until January 25, 1888. In 1879 F. Wilke opened the San Antonio Hotel, but it only operated for a couple of years. San Antonio then became almost a ghost town, with only an occasional traveler stopping at the station. The owner of the station and half a handful of others remained at the site. The Tate and Wallace Belmont–Sodaville stage line still ran, but Belmont was in a downslide, and by 1890 all stage lines had stopped running through San Antonio.

The small town came to life again in 1896, when new strikes in Esmeralda and Nye Counties led to an increased amount of stage travel. The post office, with Elsie Court as postmaster, reopened on April 8, 1896, and San Antonio was back on the map. Initial strikes at Tonopah and Goldfield in the early 1900s created a heavy flow of travelers through the district until train tracks were built leading to those booming towns, ending stage travel almost overnight. Adding to the demise of stage travel was the arrival of the automobile. The station and town struggled on for a few more years, but when the post office closed on July 14, 1906, the end was in sight.

Soon only the stationmaster was left in San Antonio. By 1910 even he had

left. The site remained part of the San Antonio Ranch until it too was abandoned in the 1960s. Time and the harsh desert sun have not been kind to the spot. All that remains of the once-impressive stage station are crumbling walls. Part of the station burned down, and without a roof to protect them, the rest of the remains decayed very rapidly. A number of smaller ruins from the business district are scattered around the old stage station. The site is marked by two large, very old poplar trees, the only green in the whole valley. San Antonio lends itself to interesting photos and exploration.

San Juan

DIRECTIONS: *From Ione, head north on Nevada 844 for 24.6 miles. Exit right and follow the road for 4 miles to San Juan.*

The same group of prospectors who, only weeks earlier, had discovered rich ore in nearby Washington Canyon came across silver ore at the head of San Juan Canyon in 1862. A camp of fifty people quickly materialized, but the veins were extremely shallow and lasted only a month before fading out. The camp folded, and most residents moved to Washington, where the situation wasn't much better. From the turn of the century through the 1920s William Easton and J.B. Dunston worked the mines in San Juan Canyon but made only a small profit.

The canyon remained quiet until the 1930s, when fairly intensive operations started. The original site of the activity was the St. Elena claim group, which had first been filed on in 1872. Clara Williams owned the claim. She hired a small crew that worked a new mine to a depth of 110 feet with over 500 feet of lateral work. But the mine yielded nothing of value, and Williams soon gave up. The second group to take part in the activity was S.H. Linka's Bi-Metallic group. It consisted of three different claims: the Bi-Metallic, the McIntyre, and the Tiger. No one ever worked the Bi-Metallic extensively. The McIntyre, which consisted of the St. Louis, the Richmond, and the Henry George shafts, was also a financial flop. The claim's only producer was the Tiger, which had shafts 30 and 40 feet deep. After producing only about $500 worth of ore the mine ran out, and the Bi-Metallic group ceased to exist.

The only mine that yielded a decent amount of ore during this revival was the Grand View, which the Market brothers owned. Work on it began in 1947, and soon the mine was more than 100 feet deep. The Market brothers built a large stone cabin to house the six men working the mine. The mine produced silver worth more than $11,000 before closing in 1950.

Just the large stone cabin is left in San Juan Canyon. Nothing at all remains of the earlier operations. Only the well-weathered mine head shafts show that mining ever took place here.

Shamrock

(Camp Tungsten)

DIRECTIONS: From Ione, take the dirt road directly south from town and follow it for ½ mile. Take a left at the fork and follow the road for 1½ miles to Shamrock.

P.A. Havens made initial discoveries of rich gold, silver, copper, and lead deposits in Shamrock Canyon in 1863. It was not until 1867, however, that prospectors starting working a few fairly large mines in the area. The Indianapolis and the North Star mines, the latter of which was also known as the Phillips, were 200 and 300 feet deep, respectively. In August 1884 the hoisting works of the Indianapolis mine burned, and operations shut down. W.S. Gage had been running the mine. The Knickerbocker Mining Company reopened the mine in 1887 but sold it to Alfred Phillips and Neil Carmichael in 1888. Production was erratic, and by the turn of the century everyone had given up working the mines.

Mercury deposits discovered by J.L. Workman in 1907 gave nearby Ione a big boost and led to the first major development of the canyon. In June

1911 the Mercury Mining Company completed a $10,000, ten-ton furnace, and the Shoshone Quicksilver Mining Company also began operations. In May 1914 the Nevada Cinnabar Company completed a new mill that started out making eight flasks a day but within a few months was making sixty. A camp, called Camp Tungsten, formed in the canyon, and the company built a boardinghouse and cabins to house the miners. The Shamrock Mines Company, which was based in Salt Lake City and which had mine offices in Ione, bought twenty-one claims in the canyon in 1923, and activity in Shamrock became more consistent. The company was incorporated in 1923 with L.E. Elggren as president and John Bluth as secretary.

The company built a small mill in 1923, and operations began. A tunnel shaft revealed that the ore carried values in both silver and gold. By 1926 the tunnel was more than 1,000 feet long. Work began on branching tunnels to reach veins discovered during the early 1920s. The company remodeled the 1923 mill to use the Vandercook mercuric cyanide system. The mill had a seventy-five-ton capacity, which was sometimes was not large enough. The overflow ore was sent to the Pioneer mill in Ione.

The main operation at Shamrock during the next few years was the tunnel mine. By 1928 the tunnel was more than 1,700 feet long. The company discovered another low-grade cinnabar deposit in 1929, but the ore was not rich enough for extended production. In 1929 the mill changed again, this time

The prosperous Nevada Cinnibar Mine was in full swing in 1906. The small camp of Shamrock is in the foreground and also extended to the left. (Central Nevada Historical Society)

from a cyanide plant to a flotation system. Activity in the district faded in late 1929. By 1930 only leasers were working the area. An earthquake in 1932 collapsed the shafts, putting an end to any thoughts there may have been of reopening the mines. During the 1970s and 1980s the Iron Mercury Mining Company worked cinnabar deposits in the canyon. The company operated off and on for a little more than ten years.

Ruins are scarce in Shamrock Canyon, for the area never developed into a real camp. Most workers elected to reside in nearby Ione. Those cabins that were built are all gone now. The Iron Mercury company used the mill, which therefore still stands. Otherwise, there is nothing of much interest in the canyon.

Tonogold

DIRECTIONS: Tonogold is located 4 miles north of Tonopah.

Tonogold was a short-lived camp that sprang up in 1914 after A.H. Shipway and Peter Wyneken discovered gold there. They made their first shipment in July, and by September fifty men were working at Tonogold. The mining attracted the interest of Tonopah residents, and as many as 100 visitors a day drove in. In October the Tonogold Townsite Company, of which Eugene Bertram was president, platted a townsite and offered 190 lots. By that time 150 men were working on twenty different leases. The Tonopah-

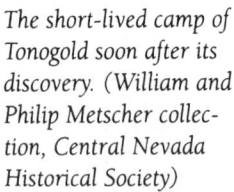

The short-lived camp of Tonogold soon after its discovery. (William and Philip Metscher collection, Central Nevada Historical Society)

Jupiter Mining Company, of which F.T. Peterson was president, formed and gained control of most of the claims. Because of the rough terrain leading to the camp, Tonopah businessmen collected donations and used the money to build a suitable road. Despite the flurry of activity, however, the ore values turned out to be poor, and the camp folded by the end of the year. After 1912 there was no more organized activity at Tonogold. Because of its short life, there were never any permanent buildings at Tonogold, and nothing is left there today.

Tonopah
(Butler)

DIRECTIONS: *Tonopah is located at the junction of US 6 and US 95, 207 miles northwest of Las Vegas.*

Tonopah Springs, which was later the site of one of the richest booms in the West, was an Indian campground for many years before Jim Butler discovered ore at Tonopah on May 19, 1900. There are a number of stories of how he discovered the ore. In the most popular of these, Butler's mule wanders away, and when he finds him he notices an outcropping that appears heavily laced with silver.

Butler firmly believed that he had happened upon a valuable silver deposit, but he had trouble convincing the assayer he visited in nearby Klondike that this was the case. Throwing the samples Butler had brought to the back of his tent, the assayer told Butler they consisted mainly of iron and were worthless. Butler was still convinced that his find was a significant one. On his way back to his Monitor Valley ranch, he stopped at Tonopah Springs once more to gather samples. Back at the ranch, he put them on his windowsill. Not much time passed before Tasker Oddie, who would later be governor of Nevada, stopped at the ranch and caught a glimpse of the ore samples. He offered to pay for another assay, and Butler agreed. Butler, in turn, offered Oddie a quarter interest of the assay, and Oddie eagerly accepted. Oddie took the ore samples to William Gayhart, an Austin assayer, and offered him a quarter interest in his quarter. Gayhart found that the assay ran as high as $600 a ton. When Gayhart notified Oddie of the value of the samples, he immediately sent an Indian runner to Butler's ranch to tell him about the rich find. Butler reacted calmly. He stayed at his ranch to complete the hay harvest and did not even bother to file claims on the lode site.

News of the discovery traveled to Klondike, and soon scores of eager prospectors were searching around Tonopah Springs, to no avail, for Butler's lode. Butler finally went to Belmont, and on August 27, 1900, he and his wife filed on eight claims near the springs. Six of these—the Desert Queen, the Burro, the Valley View, the Silver Top, the Buckboard, and the Mizpah—turned out to be some of the biggest producers the state has ever had.

Work began on the Mizpah mine in October 1900, and a camp called Butler formed nearby. On Christmas Day 1900 fourteen men were living in the camp, including Butler and Tasker Oddie, Nye County's new district attorney. Butler decided to lease out all his claims from December 1900 to December 1901. Soon the cry of "Jim, how about a lease?" rang throughout the bustling camp. Oddie and Butler were partners, receiving a 25 percent royalty on all gold and silver mined from the Butler claims.

The town of Butler began to grow by leaps and bounds. By January 1901 there were 40 men in the camp. The first stagecoach, coming from Sodaville, arrived in Butler on March 24, 1901, with seven passengers aboard. The trip included an overnight stay at Crow Springs and took two days. At that time the camp consisted of seven shacks and a number of tents and had a population of 60. Within weeks the number of people at the camp had increased to 250. A post office, with Willie Sinclair as postmaster, opened at the booming camp on April 10, 1901. The office was named Butler. It was not until March 3, 1905, that its name changed to Tonopah.

By the summer of 1901 Butler was beginning to make its mark on Nevada's silver production figures. The mines around the town produced almost $750,000 worth of gold and silver in 1901, and for the next forty years the Tonopah mines were consistent producers. The town now had six saloons, restaurants, assay offices, lodging houses, a number of doctors, lawyers, and a population of 650 that was rapidly swelling. The first wedding in Tonopah took place on November 14, when Harry Stimler and Eleanor Whitford were married. A newspaper came to the town on June 15, 1901, when W.W. Booth, who had published a paper in Belmont, set up the *Tonopah Bonanza*. The first issue included this greeting: "With this issue, the *Tonopah Bonanza* glides down the typographical ways and into the sea of journalism. Whether its voyage will be a calm and prosperous one, time alone will tell. The *Bonanza* will at all times act as a free lance, giving credit whenever merited and censure when called for. Our policy in politics will be for the best of the country. That the paper will meet with public favor or condemnation is left to the opinion of the reader and advertiser. We have done our best and sincerely hope it will meet with your approval." Until March 1905 the paper listed Butler as its place of publication. Booth took over the postmaster duties from Sinclair and served until 1905.

1902 was another extremely prosperous year for the booming town. Jim Butler had sold the claims, which were consolidated into a new company, the Tonopah Mining Company. The company was incorporated in Delaware and had stock listed on both the Philadelphia and the San Francisco exchanges. The company, with J.H. Whiteman as president, controlled 160 acres of mineral-bearing ground around the Tonopah district. It also had holdings in

A post card of the Tonopah-Goldfield Railroad station in the 1920s. The depot was the victim of arson in August 1980. The Central Nevada Historical Society had discussed using the building as a museum and later used the huge metal safe doors for the archives vault at the Central Nevada Museum. (Shawn Hall collection)

the Tonopah-Goldfield Railroad and controlled mining companies in Colorado, Canada, California, and Nicaragua. The mine workings at Tonopah consisted of three deep shafts with more than forty-six miles of lateral workings. The deepest of the three shafts was 1,500 feet. The company shipped ore mined at the site to the town of Millers in Esmeralda County, where it was treated in a 100-stamp mill. The company's mines used this mill until suitable treatment facilities were built at Tonopah.

The Tonopah-Belmont Mining Company also formed in 1902. The company, whose president was C. A. Heller, was based in New Jersey. Its property, which consisted of eleven claims covering more than 160 acres, was on the east side of the Tonopah Mining Company's property, The former company's property included two vertical shafts, one 1,200 feet deep and the other 1,700 feet deep, and had workings covering almost thirty-nine miles. Like the Tonopah Mining Company, this company also had to ship its ore to Millers until 1912, when it built its own sixty-stamp mill at Tonopah. The mill had a capacity of 500 tons. From 1912 to 1923, its years of activity, it was regarded as one of the country's best-equipped and most efficient silver cyanide mills.

These two mining companies provided the financial foundation needed to boost Tonopah to a position of prominence in 1902. There were some dark days for Tonopah that year. A month-long epidemic plagued the town.

Before it subsided, fifty residents had died. By the end of 1902 the town had recovered, and the population stood at more than 3,000. The booming town supported more than thirty saloons, a few churches, a school, two newspapers, and numerous other business establishments. Stages from all over the state began to arrive in Tonopah. One was held up on the outskirts of town. This was Tonopah's only stage robbery. Wyatt Earp was a resident of Tonopah from 1902 to 1904, running the Northern Saloon and helping the law out every once in a while.

Any fear that Tonopah would be the site of just one more quick boom vanished in late 1902 and early 1903, when prospectors located substantial new ore deposits deep in a number of the already active mines, including the Montana-Tonopah, the Desert Queen, the North Star, and the Tonopah Extension. The huge volume of ore that Tonopah's rich mines were producing created shipping problems, for at that time the only mills in the vicinity were in Millers. Construction began on a narrow gauge railroad in late 1903. The sixty-mile railroad connected Tonopah with the Carson and Colorado branch of the Southern Pacific Railroad at the Sodaville junction. The railroad was officially opened on July 25, 1904, and three days of jubilant celebration followed the opening. In 1905 the railroad expanded to standard gauge. When the rails arrived in Goldfield in the fall of 1905 the railroad was organized and named the Tonopah and Goldfield Railroad.

In May 1905 Tonopah became the county seat, taking the place of the now almost empty Belmont. The town continued to grow. John Brock dominated most of the mine ownership in Tonopah. In addition to controlling a number of prosperous mining companies, he was also acting president of the Bullfrog-Goldfield Railroad and the Tonopah and Goldfield Railroad.

Two major mining companies achieved prominence in Tonopah during the next few years. The Tonopah Extension Mining Company had formed in Arizona and had its main offices in New York City. It controlled almost 700 acres of mineral-bearing ground in the Tonopah district, most of it on the west side of the Tonopah Mining Company's property. The company owned three major shafts, which had a total of about thirty-three miles of lateral work. The three mines—the No. 2, the Victor, and the McKane—were all more than 1,500 feet deep. The deepest, the Victor, was almost 2,200 feet. In 1910 the company built a thirty-stamp cyanide mill, which it later enlarged to fifty stamps.

The second important new company in Tonopah was the West End Consolidated Mining Company, which was incorporated in Arizona. West End controlled 185 acres on the southwest side of the Tonopah Mining Company's property. The company also controlled the Halifax-Tonopah Mining Company and leased land from Jim Butler. There were three main shafts on the company's property, the deepest of which was more than 1,400 feet deep. In 1911 the company built a cyanide mill that had a daily capacity of more

The old Tonopah High School. (Southerland Studios, Carson City)

High School, Tonopah, Nev.

than 200 tons. In addition to these four big companies, nearly twenty more well-established mining companies were active in the district from the 1910s through the 1920s.

By 1907 Tonopah had become a full-fledged city, with modern hotels, electric and water companies, five banks, schools, and hundreds of other buildings. Newspapers began to play a more important role in the town, and a number of rivalries arose between them. The most notorious rivalry was the one between the *Tonopah Sun* and the *Tonopah Daily Bonanza*. The *Sun* had begun in 1904 as a weekly but became a daily on January 10, 1905. W. W. Booth, who ran the weekly *Tonopah Bonanza,* which had folded on Christmas Day 1909, also ran the *Daily Bonanza*. Booth's daily began publication on October 24, 1906, and soon the *Daily Bonanza* and the *Sun* were fighting tooth and nail over subscribers in a race to become the most prominent newspaper in the area. During this period, three other weeklies were being published.

The Mizpah Hotel, the landmark of present-day Tonopah, was completed in 1908, and it opened with great fanfare on November 17. The owners spent more than $200,000 on the hotel, which boasted baths, steam heat, and crude elevators. Not everything glittered in Tonopah, however. A fire in May 1908 destroyed an entire block of the business district. Losses totaled $150,000. Another fire in June 1909 burned the Tonopah and Goldfield Railroad's roundhouse and machine shops. Once again, losses amounted to $150,000.

February 28, 1911, was by far the darkest day in the town's mostly bright history. A mysterious fire, small but smoky, broke out at the bottom of the 1,200-foot shaft of the Belmont mine at seven o'clock in the morning. The fumes were extremely toxic, but a number of men had already gone below before anyone realized how serious the situation was. Calls to the hoist operator came in slowly at first but soon became frantic. Although many men were brought to safety, seventeen perished in the mine. All of them died as a result of the choking fumes, not from the fire. A number of the dead had actually made it to the hoist lift but were so weak that they fell out of the cage to their deaths at the bottom of the shaft. Actual damage to the shaft amounted to only $5,000. The cause of the deadly fire was a candle left on a pile of dry timber by a careless night watchman. This was the only major mining accident that occurred in the Tonopah mines. The Belmont mine soon reopened. It produced $38 million worth of silver and gold before another fire in 1939 closed it for good.

W.W. Booth remained a controversial figure as Tonopah continued to progress. In 1916 he was brought to court on a libel suit after accusing a district attorney of dishonesty. He lost the case and was sentenced to six months in the county jail. The other newspaper editors were outraged, and their voices were heard. Booth was released after serving only a month and

was eventually pardoned. His *Daily Bonanza* came upon hard times in the 1920s. On November 16, 1929, Frank Garside, who had been in charge of the *Tonopah Daily Times* since its launch on December 1, 1915, bought the paper. Garside consolidated the two papers into the *Tonopah Times-Bonanza*. Booth soon left Tonopah and moved to Hawthorne, where he became editor of the *Hawthorne News*. The *Tonopah Times-Bonanza* was a daily until April 2, 1943, when it became a weekly, as it still is today.

Tonopah's mines continued to produce extremely well until the Depression brought a slowdown. From 1900 to 1921, the district's peak years, they produced ore worth almost $121 million. Tonopah's biggest year was 1913, when its mines yielded almost $10 million worth of gold, silver, copper, and lead.

By the time World War II started, only four major mining companies were operating in Tonopah. A huge fire in October 1942 destroyed the Tonopah Extension mill and property and spread to a nearby hotel, causing $100,000 worth of damage. At the end of the war even the companies that had been there at the beginning were gone. The final blow came in 1947 when the Tonopah and Goldfield Railroad folded and its rails were torn up.

Not much mining activity has taken place in Tonopah since then. In 1968 Howard Hughes and his Summa Corporation bought 100 claims in Tonopah, including the Mizpah, the Silver Top, and the Desert Queen mines. Hopes for a mining revival soon faded, however, after the company took core samples and nothing more. A few of the old mines were re-timbered, but they never reopened. As of now, the value of the Tonopah district's total production is just over $150 million, a figure few other mining towns can top.

Tonopah had expanded greatly by the end of 1901. (Philip Metscher collection, Central Nevada Historical Society)

Tonopah is still the county seat of Nye County and has a population of around 2,500. In recent times, the town's proximity to the Nellis Air Force Bombing and Gunnery Range has allowed many businesses to remain there. A fair number of military personnel lived in the motels while working at the missile test range. A part of the test range that is close to Tonopah was home to the super-secret Stealth fighters and bombers. Long before the military acknowledged their existence, the planes frequently treated residents of Tonopah to flyovers. But the Stealth base has been relocated to New Mexico, and other cutbacks have drastically reduced the military's impact on the town.

Tonopah is still vibrant. Tourism now plays a large part in the local economy. Many buildings from the boom era still remain. These include the Mizpah Hotel, which was completely renovated in the late 1970s. Other points of interest in the town include the Nye County Courthouse, built in 1905 at a cost of $55,000 on land donated by Jim Butler; the old Tonopah Public Library, built in 1912 and now the oldest active library in Nevada; St. Mark's Episcopal Church, built in 1906 and one of the better-known landmarks in Tonopah; and the ruins of the old 500-ton Belmont mill, on the east side of Mount Oddie. A must for any visitor is the fantastic Central Nevada Museum, which features many wonderful displays and a complete research section. This author, however, found the most interesting part of the museum to be the extensive outdoor displays, which include large and small artifacts from all over the area. The artifacts include parts of a stamp mill from Manhattan, hoist cages, parts from military aircraft that crashed during Tonopah's years as a military base, buildings from various towns, and a nature walk featuring native plants and fauna. The museum was originally intended to be housed in the old Tonopah depot, but the depot burned in August 1980. Only the heavy vault doors needed to protect the silver and gold bars were saved, and they now adorn the archives vault at the museum. Another interesting place to visit in Tonopah is the Tonopah Mining Park. Echo Bay Mines gave the property to the town, and a walking tour of the mining complex has been organized. The walk features numerous old head frames, buildings, mining equipment, and a huge glory hole.

Tonopah is an excellent home base for anyone visiting ghost towns in Nye County. Supplies of all sorts and reasonably priced lodging are available there.

Toyah

DIRECTIONS: *From Ione, head east on Nevada 844 for 8.25 miles. Exit right and follow the road for 4.25 miles. Exit right again and follow the road for 3½ miles to Toyah.*

Toyah, a stop on the Belmont–Ione stage run in the 1860s, was located at the highest point of Ophir Canyon. The station served as a rest stop for the horses after the tough climb up the mountain. Only a large wooden cabin and a small stable were ever built there. Later, very limited mining took place near the site.

Hardly a trace remains of Toyah. Some scattered test tailing piles mark the general area along the road where the station once stood. Nearby are some very faint wood remains that are difficult to see. This is all that is left of Toyah.

Union

DIRECTIONS: *Union is located 1 mile east of Berlin and is well marked.*

Union, a minor camp, sprang up in Union Canyon in 1863, and the Union Mining District formed on May 30, 1864. The Atlantic and Pacific Mining Company owned all the mines in the district and platted a townsite in

1865. Under the guidance of George Veatch, the company began construction of a twenty-stamp mill. However, the company ran into financial problems and folded in 1866 before the mill was completed. The camp faded quickly and was virtually nonexistent during the 1870s and 1880s.

Union revived when prospectors discovered rich gold deposits in the hills around Berlin. Some of the miners working in the Berlin mine settled in Union Canyon because it was far enough from the mine and mill that it was quiet but close enough to enable them to walk to work. But the canyon's peace was about to be disrupted. In May 1895 John Mayette began construction of a ten-stamp mill in Union. He obtained the stamp battery from Belmont and the mill building from Ellsworth. At the same time, Theodore Cirac was building his own ten-stamp mill in the area. Mayette's mill started up in July, and Cirac's started up in November. On January 20, 1896, a post office, with Theodore Cirac as postmaster, came into being, but it closed on July 25.

The little settlement reached the peak of its activity in 1904 and 1905. During this period it contained more than twenty buildings, including a town hall, a schoolhouse, a mercantile store, and the two mills. The town hall was a large, one-room log structure. In 1910 Bob Dixon moved the building to Berlin, where he used it as a combination bar and boardinghouse. The schoolhouse was just a small log cabin. In 1902 enrollment stood at twelve, and Celia Peter was teacher. Peter lived in what now is the only building left in Union. Mr. Lennox ran the mercantile store, which was also a saloon. It closed in 1910, when the miners' strike in Berlin emptied the town. The Cirac family, all of whom resided in the canyon, owned the Cirac mill. Louie, one member of the family, owned the most imposing structure in Union, a two-story brick building with a large room on the second floor that frequently served as a dance hall. Louis and three of his brothers later became the first to patent the stop-and-hold automobile lock, making them the most notable people ever to have lived in Union. The town also had another small saloon located next to the Cirac mill. It belonged to a Chinese man known as Sam "China" Wing.

Another mill, the Mayette mill, was situated at the mouth of Union Canyon. The Mayettes' house in the canyon was surrounded by fruit trees and a large garden. A nearby spring supplied it with running water. Ruins of the mill still exist today.

The camp at Union was active until late 1910, when miners went on strike at the Berlin mine. The company refused to grant them a pay hike and closed down the mine. Both Berlin and Union soon emptied. The years have not been kind to Union. The site is almost completely obliterated. Other than the still-standing adobe house, the only ruins marking the site are one partial adobe wall, which used to be part of the Kennedy house; the concrete foundations of the Cirac mill; and some scattered lumber north of the adobe house. Broken glass and rusted tin cans mark the site of the Lennox mercantile store, and broken beer bottles and a dugout shelter are all that remain of Sam Wing's saloon. Union is now part of the Berlin-Ichthyosaur State Park. The site is patrolled by rangers who try to protect what has not yet been destroyed in Union from further vandalism.

Washington

DIRECTIONS: *From Ione, head north on old Nevada 21 for 32 miles. Exit right and follow the road, keeping to the right, for 10 miles to Washington.*

Prospectors' original discoveries in 1860 in Washington Canyon were later developed as the Warner mine. Further discoveries in 1862 and 1863 led to a fairly active rush to the area. A townsite was platted in 1863, and soon Washington had three saloons, a bakery, a livery stable, two stores, and the first billiard hall in Nye County.

The Washington, or Columbus, mining district formed in 1863. The New Hope Mining Company built a $40,000, ten-stamp mill in the same year. The camp also served as a stop on Thomas Morgan's Austin–Ione stage line. The rush was over by late 1864, and the stamp mill closed when the supply of ore came to an end. Many of the mines were not developed until the early 1870s after capitalists began to invest in the Washington mines. A number of miners continued to work a few mines. Active mines included the Pyrennes, the Revenue, the Valparaiso, the Stranger, the Golden Fleece, the La Chili, the Santo Nino, the St. Helena, the New Orleans, and the Mobile. About thirty men worked these mines. By 1865, however, the mines had all closed down.

The remaining inhabitants of Washington Canyon turned to agriculture for a living. The canyon's rich soil was excellent for root vegetables and cabbage, which had markets in Austin and Ione. Because of the abundance of trees in the canyon residents altered the stamp mill to make it a sawmill. The mill enjoyed a long production period, during which the sides of Washington Canyon were denuded.

A small-scale silver mining revival took place in Washington Canyon in 1870. A post office, with Barney McGirr as postmaster, opened on July 29, 1870, but only lasted until August 27, 1872. After the mines closed again in late 1872 only a handful of people remained in the canyon. By the beginning of the 1880s the entire canyon belonged to ghosts. In November 1912 the newly formed Washington Mining Company bought twenty-two claims in the canyon for $4 million, $3 million of which went to Rose Weiner. But the sellers of the claims received hardly anything, because the company folded before any production took place.

The canyon remained empty until 1918, when the Warner Mining and Milling Company reopened the Warner mine and began new operations. A. O. Jacobson and A. P. Swoboda, both from Salt Lake City, ran the company. The pair were also officers in the Columbus-Rexall Mining Company and the West Toledo Mining Company. In addition, Jacobson ran his own mining company, the Jacobson Mining and Milling Company, which was active elsewhere in Nevada. The Warner Mining and Milling Company purchased twenty-one claims covering more than 400 acres. The owner of the company was Rose Warner, one of a very small number of female mine owners. The claims included three tunnel mines and the Warner mine shaft. The ore from the district was mainly quartz and carried high values in silver, with smaller amounts of lead and copper. The company worked the Warner mine from the old depth of 100 feet to more than 600 feet and installed a diesel engine at the mine. It built a new cyanide and concentration plant that used a crusher, rolls, tables, and cyanide tanks. A 1,500-foot aerial tram ran from the Warner mine to the mill. The company continued to operate the mine until 1922, when financial backing was withdrawn and it folded. After that, only cool breezes visited the canyon.

The district reopened in the early 1930s when a small-scale revival took place. Another small mill was constructed, along with a tram that ferried the ore from the mine high up on the canyon to the mill on Washington Creek. This bout of activity was short, and by 1937 the district was once again abandoned. Miners worked a small tungsten mine from 1956 to 1957, and then Washington became ghostly again.

The ruins of the three mills remain in the canyon. There are also a number of stone ruins left from the activity that took place in the 1860s and 1870s. Although the remains are scant, the site is extremely interesting and is well worth the long trip. This is a great place for the ghost-town enthusiast to enjoy uninhibited exploration.

Willow Springs

DIRECTIONS: *Willow Springs is located 4 miles northwest of Cloverdale.*

Willow Springs was established in May 1867 as a stop on the Wells Fargo Belmont-Westgate stage line. A small, crude station was built there. However, the site was only used for a short while. Once nearby Cloverdale became established, the Willow Springs station was closed. Only some rocks and the spring mark the site today.

Northeastern Nye County

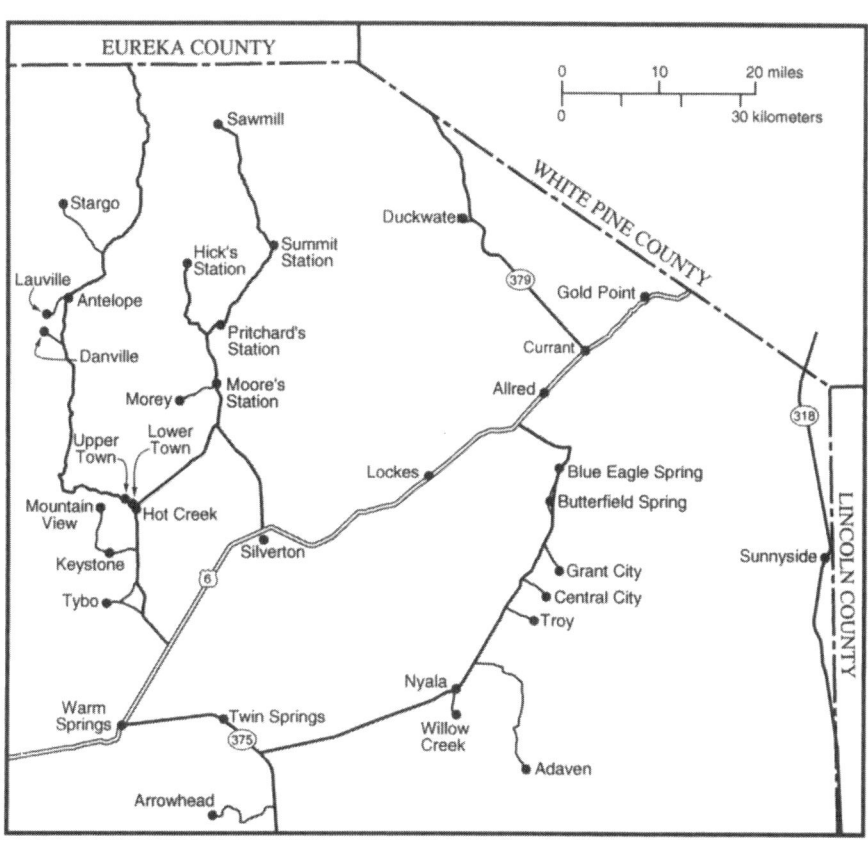

EUREKA COUNTY

0   10   20 miles
0   30 kilometers

WHITE PINE COUNTY

Sawmill

Stargo

Duckwater

Hick's Station

Summit Station

Lauville

Antelope

379

Gold Point

Pritchard's Station

Currant

Danville

Moore's Station

Allred

Morey

318

Upper Town

Lower Town

Lockes

Blue Eagle Spring

Butterfield Spring

Mountain View

Hot Creek

Keystone

Silverton

6

Grant City

Central City

Sunnyside

LINCOLN COUNTY

Tybo

Troy

Nyala

Warm Springs

Twin Springs

Willow Creek

375

Adaven

Arrowhead

Adaven

(Sharp)

DIRECTIONS: *From Currant, take US 6 for 9.8 miles. Exit left and follow the road for 6 miles. Exit right and follow the road for 24 miles. Exit left and continue on this road for 15 miles to Adaven.*

Adaven was originally known as Sharp, after Thomas Sharp, a rancher who settled there in the 1870s. A small settlement formed around his ranch, and the Sharp post office opened on December 14, 1901, with Lewis Sharp serving as postmaster. The population remained at around twenty-five for the next fifty years. On May 1, 1939, the town was renamed Adaven (Nevada spelled backwards). The Adaven post office continued to operate until November 30, 1953. There are still small ranches near Adaven, and a number of older buildings remain. The main ranch belongs to the Sharp family to this day.

Allred

DIRECTIONS: *From Currant, head south on US 6 for 5 miles to Allred.*

Allred boasted a small post office from April 17, 1911, until October 31, 1912. Allen Oxborrow and George Kump served as postmasters. Very little is known about the site, and nothing remains there.

Antelope

(Clear Creek) (Crockers Ranch)

DIRECTIONS: *From Hot Creek, continue west through Hot Creek Canyon for 6½ miles. Exit right and follow the road straight through for 20 miles to Antelope.*

Antelope was a small stopping place in Little Fish Lake Valley in the late 1800s. A little ranch had been active there since the 1870s, but it was not until mining activity in the valley picked up that a regular settlement began to form. A post office, with Orange Wattles as postmaster, opened on February 8, 1904. Antelope faded after activity at nearby Danville, Stargo, and Lauville ended. The post office closed on July 31, 1908, and soon the Antelope complex returned to ranching.

In 1944 miners worked a small, open clay pit just to the west of the old ranch. The operation, run by the Coen Companies, Inc., of Los Angeles, employed four men and only lasted for a short while. The ranch was aban-

This 1860s cabin
struggles to survive.
(Shawn Hall collection,
Nevada State Museum)

doned soon after that, and the site has slowly deteriorated. Today, a number of buildings remain at the site. The main ranch house is still used to house cowboys when they work cattle in the valley. One small boardinghouse, a blacksmith shop, and a few old branch-and-clay stables are among the other interesting remains. A number of small dugout stone cabins are located on the side of a small hill about 200 feet from the main cluster of buildings. The site is definitely worth the trip.

Arrowhead

(Needles)

DIRECTIONS: *From Warm Springs, head east on Nevada 375 for 19½ miles. Exit right on Reveille Road and follow it for 5½ miles. Exit right again and follow this unimproved road for 3½ miles to Arrowhead.*

The Arrowhead district was organized in August 1919 when prospectors discovered substantial silver deposits in the Reveille range. The camp grew rapidly and soon contained a small group of buildings. A post office, with William McDonald as postmaster, opened on December 24, 1919. In February 1920 Nevada Wireless Telegraph and Telephone opened a station at the camp, and W.W. Haley opened the two-story Arrowhead Hotel, which he had moved from Goldfield.

Many different mining companies worked the Arrowhead district during its short existence. Seven of these were established during the abortive boom

of 1919. The two largest companies were the Arrowhead Extension Mining Company and the Arrowhead Mining Company, the latter of which owned the Arrowhead mine.

The Arrowhead Extension Mining Company, of which John Kendall was president, was based in Tonopah. M.J. McVeigh was vice president and general manager of operations at Arrowhead. The company was incorporated in 1919 with $150,000 in capital, and shares were offered to the public. The mining company purchased 100 acres south and east of the Arrowhead mine, and a shaft was sunk. The extracted ore brought $65 to $70 a ton, and the company dug the shaft to a depth of 150 feet before the mine closed.

The area's other important company, the Arrowhead Mining Company, was also based in Tonopah. R.L. Johns was president. Incorporated in September 1919 with $150,000 in capital, Arrowhead Mining purchased four claims, one of which it eventually developed into the Arrowhead mine. The ore from that mine was quite rich, assaying from $100 to $200 a ton, but by 1921 its value had dropped to $25 a ton. Construction of a small mill was under way at the time, but the mine closed before the mill was completed. Developments in the mine were extensive, and before activity ceased its vertical shaft was over 350 feet long.

The remaining mining companies produced limited quantities of ore. The Arrowhead Annex Mining Company, of which J.W. Gilfoyle was president and which was established in 1919, purchased four claims north of the Arrowhead mine. A shaft was sunk, but it was never very successful, and the company soon folded. The Arrowhead Wonder Mines Company was also active in the district. Based in San Francisco, the company, of which W. H. French was president, had seven claims northeast of Arrowhead, and it dug two shafts. For awhile the company was extracting ore running $110 per ton, but the veins in the Wonder mine, which miners believed were the same as those being worked in the Arrowhead mine, soon ran out.

The only other company of any consequence in the district was the Arrowhead Syndicate Mines Company, which was based in Ely. C. L. Osterlund was both its president and its manager. The company formed in 1919 and sold 480,000 treasury stock shares, the proceeds of which it used to purchase eighty acres south of the Arrowhead Mining Company's property. With the help of an eighteen-horsepower hoist, the company dug a shaft to a depth of 200 feet. But before it was really able to get moving the company ran out of money, and it was forced to curtail operations in August 1920. Hopes of reopening in 1922 vanished when the company was unable to raise the necessary capital.

Of the other companies that worked the area (the Arrowhead Bonanza Mining Company, the Arrowhead Esperanza Mines Company, the Arrowhead Wonder Mining Company, the Arrowhead Signal Silver Mining Company, the Arrowhead Inspiration Mines Company, and the Arrowhead Consolidated Mining Company), only Arrowhead Consolidated's mines produced substantial amounts of ore. The district's first ore shipments were not made until March 1920. In January 1921 R. L. Johns bought the Blue Jay claims—the first locations—from Frank Schultz and then formed the Silver Arrowhead Mining Company.

The Arrowhead district's active period was fairly brief, and by 1922 the town was declining rapidly. The post office closed on September 4, 1924. Other postmasters who served were Joseph Ehrlich and Ethel Allred. Ehrlich's nickname was Whispering Joe, because he had lost his voice in an avalanche years before. The town was completely abandoned shortly after the post office's closure, although in June 1926 the Arrowhead Development Company, of which Peter Fox was president, formed and reopened some of the mines. The company shipped ore to the company mill in San Bernardino, but values dropped, and operations ceased in 1928. While leaching operations took place at nearby Reveille in the 1980s, in recent years no one has shown much interest in Arrowhead, which continues to slumber. The only building remaining at Arrowhead is the blacksmith shop. Extensive mine dumps left by the many companies that once worked Arrowhead are scattered throughout the surrounding hills. Concrete foundations and some rubble are the only other remains that mark the site.

Branding at the Blue Eagle Ranch during the late 1880s. (Nevada Historical Society)

Blue Eagle Spring

DIRECTIONS: From Currant, take US 6 south for 9.8 miles. Exit left and follow the road for 6 miles. Exit right (south) and follow the road for 3 miles to Blue Eagle Spring, located on the right side of the road.

Alexander Beaty, Francis Tagliabus, Henry DeGroot, Ernest Hahn, and F. B. Clark originally homesteaded Blue Eagle Spring, in Railroad Valley, in the 1860s. Martin "Pop" Horton built a ranch house there. He also constructed a store, a blacksmith shop, and a saloon and ran a station house that served many tired travelers. The Blue Eagle Spring Ranch became the center of the Blue Eagle ranching district. Nevada Governor Jewett Adams bought the ranch in 1890. In 1896 he sold out to Lewis Sharp. The store at the ranch remained open until 1910. A school operated there until 1960. The ranch has continued to operate to this day. The Sharp family ran the ranch until 1965 when it was sold to Carole and Carl Hanks. Carole is the daughter of Lina Sharp. A few original buildings are left on the ranch, including the old store and the saloon. In 1954 oil began to be produced nearby. Texota Oil Company drilled the first flowing oil well in Nevada in the field in 1963. A refinery built in 1966 burned in October 1968. Since then, Blue Eagle Spring's oil has been shipped to Bakersfield and Salt Lake City. The oil fields of Railroad Valley are the best producers of all the fields in Nevada.

Butterfield Spring
(Haystacks)

DIRECTIONS: From Currant, head south on US 6 for 9.8 miles. Exit left and follow the road for 6 miles. Exit right and follow the road for 6 miles to Butterfield Spring.

Butterfield Spring has been a source of cool, refreshing water for centuries. The Indians were the first to use the site, making it a camping spot during their hunting trips. It was not until 1867 that whites came to the spring. Henry Butterfield, who had a salt mine ten miles south, was the first of them to arrive. In 1871, after Butterfield left, Alexander Beaty, who later made important ore discoveries at Troy, decided to homestead the land and soon built a small ranch complex there. He sold out to George Sharp in 1895 for $350.

Nearby Butterfield Marsh had large deposits of sodium chloride that was

Only one building stands next to Butterfield Spring. (Shawn Hall collection, Nevada State Museum)

almost 99 percent pure. Much of this salt was mined during Tybo's boom for use in its silver mills. After the salt mining came to an end, no activity other than ranching took place at Butterfield Spring until 1912. That year, the Railroad Valley Company formed and began exploratory drilling, hoping to find potash salts. The company drilled seven test holes ranging from 745 feet to 1,200 feet but failed to find any potash salts. The company spent almost $150,000 on this project and when it failed immediately went bankrupt.

The spring itself makes Butterfield Spring unique. The depth of the spring is unknown, but its clean water stays at a constant temperature of fifty-four degrees Fahrenheit. A strain of chub, unique because of its isolation, lives in the spring. Most of the original buildings of the old ranch, including a few sod-type structures, remain. The Sharp family still owns the ranch.

Central City

(Irwin) (Seymour)

DIRECTIONS: *From Nyala, continue north on the Currant-Nyala Road for 12½ miles. Exit right and follow Irwin Canyon Road for 3 miles to Central City.*

The small camp called Central City formed in the early 1870s after prospectors from Troy and Grant City found ore in the canyon. A mining district called Seymour was soon organized. The camp quickly faded but was revived in the early 1880s when nearby Grant City also experienced a revival. Central City's population reached twenty-five at one point, but the camp was abandoned by 1883.

The canyon remained empty until 1905, when F. L. Irwin staked a number of claims there. Irwin's father, Isaac, had established the nearby Irwin Ranch in 1870. Soon a small camp of twenty formed. A five-stamp mill was completed in June 1911. While the mill existed, miners did several hundred feet of work on a high-grade gold vein. The vein gave out shortly after the mill started, and everyone soon left the canyon. In 1913 George Seay, one of the discoverers of Tombstone, bonded all Central City's mines, but he didn't have any luck and gave up in 1914. Interest in the canyon's claims was renewed in the 1930s when two eighty-foot shafts, known as the Mayolli claims, were sunk in the area. Ore from the two mines assayed from $8 to $16 per ton, and in 1940 a small mill began operation at the camp. The mill consisted of a crusher, a four-foot ball mill, and a small jig. This bout of activity did not last long, and soon the canyon was empty once more. In 1953 Bud, Starle, and Solan Terrell began working on a tungsten deposit at Mud Spring, located just north of Central City. The trio moved equipment from their claims at Eden Creek. A three-stamp mill, which was later enlarged to five stamps, proved a failure. The Terrell family maintained control of the mine until the 1980s, though little work was done on it after the 1950s.

Three small cabins still stand at the site, all dating back to the 1930s. There are also two stone ruins that appear to be from buildings erected during the earlier settlement of Central City. The road to the site is extremely rough and might not be passable.

Currant

DIRECTIONS: Currant is located on US 6, 114 miles east of Tonopah.

Currant began not as a mining camp but as a farming community. The site was first homesteaded in 1868, and within a few years ten families had moved to the area. Situated at the head of Railroad Valley, the town became the supply point for the valley's many ranches. During the 1870s Currant was the supply point for Tybo. Currant was located at the crossroads to Eureka, Hamilton, White River, and Railroad Valley. Thomas and Richard Barnes started a stage line to White River Valley in 1872 and opened a station at Currant. A post office, with Alexander McCullough as postmaster, was established at Currant on April 16, 1883. At the time, the small town had a population of around fifty and included a couple of stores, one saloon, and a number of other buildings. The post office closed on May 5, 1884, but the small settlement continued to serve the ranchers' needs. The size of Currant's population remained fairly constant throughout the rest of the century. The post office reopened on September 19, 1892. Mamie Blackwell was postmaster. The office remained open until July 31, 1922.

Mining activity came to the Currant area in 1914, when prospectors discovered gold ore on the Shepherd Ranch. Soon some small shipments had been made. Ore from the property contained not only gold but also traces of lead and copper. Miners dug a small mine, the Sunrise, in 1916, but this and the other small-scale developments in the area never became significant. When mining activity stopped, Currant wasn't affected at all.

The town continued its sleepy existence for a number of years. The post office reopened once again on August 31, 1926. Minnie Callaway was postmaster. Nothing of historical consequence happened in Currant until the late 1930s, when prospectors discovered magnesite deposits in the newly formed Currant Creek Mining District, which encompassed parts of Nye County and White Pine County. The major deposits were located in White Pine County. Nye County contained some small but highly productive claims. The first and largest of these was the Windous Group, located one quarter of a mile into White Pine County. The deposit, which Tom Windous discovered in 1939, was on land owned by a rancher named Munson. Windous purchased the land from him and then leased the property to the Westvaco Chlorine Products Corporation. The lease ran from November 1940 to February 1942. During that time the company did extensive exploration, uncovering a 500-foot by 100-foot belt of magnesite that miners worked using the glory hole method. The company dug a number of tunnels and shafts, which contained over 700 feet of workings.

Another major magnesite deposit in Currant was the Ala-Mar, formerly known as the Manzoni vein. The deposit, which was located in Nye County, was the largest in the district after the Windous. The Ala-Mar, which the Ala-Mar Magnesium Company, Inc., worked, consisted of nine claims just south of the county line. Active exploration began in 1940, and soon the ore was being shipped out. The company made a forty-ton shipment to the General Electric Company in early 1941. There was quite a bit of development at the site. The Ala-Mar Magnesium Company set up a small tram line that ran from the glory hole to the shipping building, where trucks would take the ore for shipment. It also erected three buildings, including one that served as the compressor house. But the company produced only 500 more tons of ore before folding in 1942.

Gold was also mined in the Currant district. Steve Pappas and George Bogdanovich discovered gold ore assaying at $256 a ton in April 1939 and organized the Currant Creek Mines Company. The American Smelting and Refining Company optioned some of the company's claims for $125,000 but returned the lease because the ore was of limited sustained value. In June 1940 legal problems stopped all activity when Fred Farnsworth obtained a restraining order against the company. Farnsworth claimed that he had been with Pappas and Bogdanovich when they had made their discoveries, and he wanted his share. The suit was resolved amicably when Farnsworth got a share, and the claims were immediately optioned to the Comstock Gold Point Mines Company for $500,000. But the analysis remained the same, and the claims reverted back to the Currant Creek company. The company tried to find more rich veins but was unable to interest any other companies in the venture, and the men gave up in 1942.

The Currant post office closed for the last time on December 31, 1943. Minnie Callaway, who had become postmaster in 1926, died in 1943 while serving, and the office closed shortly after her death. The school closed in 1966 and burned as a result of an arson attack that took place on Halloween in 1970. Today Currant has a population of about sixty-five. Most of the residents work the small oil wells to the south. The majority of the buildings in Currant are relatively new, and there is a small trailer park for the oil well workers. The remaining old ruins consist of a few small wood cabins. The Currant cemetery is located a couple miles north of town. The gas station and restaurant closed in 1996 because of decreased traffic.

Danville

(Chloride Mining District)

DIRECTIONS: *From Tonopah, take US 6 east for 34 miles. Exit left and follow the Little Fish Lake Valley Road north for 46 miles. Exit left and follow this rough road for 3½ miles to Danville.*

The Danville district was organized in 1866 when P. W. Mansfield discovered rich silver ore in the Monitor range. Mansfield was one of a small group of prospectors who ventured into Danville Canyon in hopes of finding a shortcut through the range. His discovery was not developed until 1870, when the district was reorganized. Miners started working five mines, and soon almost thirty miners were working the area. They established a freight line to Eureka to bring in supplies at a cost of $25 per ton.

Danville's five principal mines were the Sage Hen, the Boston, the Eucalyptus, the Argonaut, and the Richmond. In addition, prospectors made thirty other claims along the canyon. The Boston, which was the deepest shaft, had reached 150 feet by 1872. At 125 feet, the Eucalyptus mine's tunnel was the longest of the four. Miners sent initial ore shipments to Austin, but after the Morey mill was built in 1873, ore from the Danville district was shipped there for processing. Some of Danville's ore assayed as high as $600 per ton.

The district was quiet for a while after 1874, but in 1877 twenty men were still working the Danville Canyon area mines. Two mining companies, the Danville and the St. Louis, ran all the mines. Alexander Trippel was the superintendent of both companies. He later took over the same job at the Morey mines, which he oversaw until his death in 1897. John Phillips bought the mines for $18,000 in December 1877 and kept Trippel on as manager. Phillips expended $91,000 over the next four years but saw little return. He never visited the mines, relying on Trippel to do it for him. He expressed doubts (the exact nature of which are unclear) as to whether he was getting his money's worth and shut down all his operations on January 1, 1882. Most of the other Danville area mines had closed in 1879, and by 1881 only five people were left in the camp.

Prospectors made a new silver discovery further up the canyon in early 1883, and soon there were twenty-five miners living in Danville again. A post office, with Frank Miller as postmaster, opened on November 21, 1883. This suggested that the camp might be around for a while. But the revival only lasted a little over a year, and the post office closed on September 8, 1884. The ghosts reigned over Danville from 1886 to 1909, disturbed only by an occasional prospector. During the early 1890s O. S. Wattles, who owned a ranch in nearby Little Fish Lake Valley, worked the Boston mine and shipped some ore to Salt Lake City, but he made very little money.

In 1909 the Boston and Sage Hen mines both reopened. Miners periodi-

*Ruins at Danville.
(Shawn Hall collection,
Nevada State Museum)*

cally removed small amounts of ore from them until 1914. Everyone then gave up on Danville until World War II began, when the call for valuable minerals revitalized the district once again. However, the ore mined was of such a low grade that operations ceased in early 1942. The total production value of the Danville district's ore stands at about $43,000. The bulk of this ore was produced from 1944 to 1946 during a two-year revival of the district. In those two years miners extracted $27,000 worth of silver and gold from the Danville mines. Most of this activity took place on the Boston claim, which had last been worked in 1934 by the Continental Mines Company of Denver. In 1949 Jack Ekstrom shipped seventy tons from the Boston claim. When the property was abandoned in late 1949 its shaft was more than 130 feet deep. After Ekstrom departed Danville, nobody tried to mine the district again.

Danville's remains are quite scant. Since the population of the small camp never exceeded thirty-five, few substantial buildings were ever constructed. Two wooden cabins remain, and a Bureau of Land Management survey crew uses them on occasion. A few ruins are scattered along the canyon. A fairly substantial, carefully crafted stone ruin is left. The purpose of the original structures is undetermined. It appears to have been made of sandstone, and each stone is precisely cut and fitted. Although only walls remain, even they are impressive. The road to Danville is extremely rocky and has some dangerous sand traps. Exercise extreme caution.

Duckwater

DIRECTIONS: *From Currant, take Nevada 379 north for 21.6 miles to Duckwater.*

Duckwater is primarily an Indian settlement, although whites did settle there in 1868. The first white settler was Isaac "Ike" Irwin, who raised hay to sell in Hamilton. Irwin's descendants still reside in Duckwater, and structures he built still stand. Irwin died in 1893. Another early settler in the area was John Williams, whose homesteaded land would later become the Duckwater Indian Reservation. Indians occupied the site long before the Europeans came to America. A post office opened at Duckwater on January 6, 1873. A school also opened. The small town gradually became predominantly Shoshone, and by the turn of the century its population stood at more than 100, a figure that has remained fairly constant since then. The Duckwater post office closed on January 29, 1941, but reopened on May 16, 1950. The Shoshone Indian Reservation, which officially formed in 1940, now contains more than 3,700 acres. A number of older buildings still remain at Duckwater, scattered among newer buildings.

Gold Point

DIRECTIONS: *From Currant, head north on US 6 for 9 miles.*

Gold Point was not so much a town as a general ranching area north of Currant. There are a number of abandoned ranches in the area. The most impressive of these is the old Fisher Ranch, where a number of older buildings still remain. The most unusual building on the ranch is a very attractive false front, the original purpose of which is unknown. There are three small mines northwest of the Fisher Ranch, but they were all prospectors' individual efforts, and they never produced. A discussion of their history is included in the description of Currant. The Gold Point area is well worth visiting just to see the old false front.

Grant City

DIRECTIONS: *From Currant, head south on US 6 for 9.8 miles. Exit left and follow the road for 6 miles. Exit right and follow this road for 10 miles. Exit left and follow Grant Canyon Road for 3½ miles to Grant City.*

The Grant Mining District formed on October 27, 1868, after prospectors made discoveries in the Grant Range in 1867. A small townsite, called Grant City, formed soon afterwards. The town had a street system and a number of small stone cabins. By the summer of 1869 Grant City had a population of more than 100 and contained a number of businesses including a saloon, a blacksmith shop, and George Norton's assay office. Norton had some claims just to the south of Grant City. He organized the Old Colony Mining District, for which he served as recorder.

The two veins that supported the camp were the Meridian and the Blue Eagle. Their ore assayed as high as $300 per ton. A test shipment of ore sent to Austin in 1869 returned $500 to $600 per ton, but this rich lode was destined to run out quickly. By 1870 the town was sinking rapidly, and soon it was empty. A small-scale revival in the early 1880s brought about twenty-five people back to Grant City. But in 1884 the town was left to the ghosts, and only the cries of coyotes echoed through the crumbling ruins.

Grant City's remains are located amidst a small stand of trees. There are a few stone ruins left from the 1860s. Wood remains in the canyon are probably from the revival of the 1880s or from the 1940s, when some people from Los

Angeles used the area as a summer lodge. The site, while it is not large, is interesting and worth seeing. There is a small spring at Grant City, and the water is excellent.

Hick's Station

DIRECTIONS: *From Warm Springs, head north on US 6 for 20.9 miles. Exit left and follow the road for 2½ miles. Take the right fork and continue for 7 miles. At the four-way junction, take a right and continue for 5 miles. At the end of the road, take a left onto a new road and continue for 8 miles. Take the left fork and continue for 1 mile. Bear right at the fork and follow the road for 8 miles to Hick's Station.*

Hick's Station, named for owner C.W. Hicks, was a stop on the Warm Springs-Eureka stage route and the Tybo-Eureka stage route during the 1870s and early 1880s. The station was not as carefully constructed as were nearby Pritchard's and Moore's stations. The advantage of Hick's Station was that it had an abundance of water, although the Shoshone Indians called the springs *sapiaua*, which means scum water.

The station was the scene of a murder in October 1875 when Ole Johnson and Richard Mason, both employees of the Eureka and Tybo Stage Company, had a disagreement over money owed from a poker game. Johnson shot and killed Mason and was sentenced to twenty-five years at the Nevada State

This is all that remains at Hick's Station. (Shawn Hall collection, Nevada State Museum)

Prison. In November 1877 Johnson, a ringleader of a prison break, was killed during the escape attempt.

A number of buildings still remain at Hick's Station. A small ranch began operating there after the stages stopped running, but the site is clearly marked with a No Trespassing sign, so it isn't possible to examine it closely. Hick's Station's last resident was Frank Farnsworth, who ran the same ranch that is still operating there today from the 1920s until his death in 1956. A small pond at the site hosts a number of wild ducks. The last half mile leading to the site tends to be extremely treacherous and includes some very long and deep sand traps.

Hot Creek

DIRECTIONS: *From Warm Springs, take US 6 north for 20.9 miles. Exit left and follow the road for 2½ miles. Take a left at the fork and follow the road for 9 miles to Hot Creek Ranch. Remains are on the ranch property.*

The Hot Creek Mining District formed in 1866 after prospectors discovered ore in a number of canyons in the Hot Creek range. In early 1867 a small town formed along Hot Creek and was named after the creek, which had been named for the steam that often rose from its surface in the morning. The town grew very fast, and by the summer of 1867 it contained a number of small stone buildings. On August 7 the government awarded Hot Creek a post office, and Lafayette Joselyn was the first postmaster. Hot Creek peaked in 1868, when its population was over 300. Business establishments in the town included a number of saloons, a hotel, a blacksmith shop, a restaurant, and an assay office. Two stamp mills, located to the north at Carrolton and Lower Town, were built in 1867 to process ore from the Hot Creek mines.

Hot Creek began its decline in late 1868 after new strikes in White Pine County lured most of the 300 residents away. Despite this, J.T. Williams completed a $6,000 stone hotel in Hot Creek in 1875. The town continued to function at a low level until the summer of 1877, when Henry Allen contracted with the Tybo Consolidated Company to build fifteen brick kilns. The kilns' purpose was to supply charcoal for the lead smelters in Tybo. The kilns, which were built in two nearby canyons, were completed in September 1877. Their construction required the use of 600,000 bricks. The kilns were an average of 25 feet in diameter, and each had a capacity of 1,400 bushels of charcoal. In 1878 Joe Williams completed a new and larger hotel at Hot Creek, complete with a large dance hall and bathhouse.

A minor revival took place in Hot Creek in 1880, and a small, ten-stamp mill that treated ore from mines in Old Dominion and Rattlesnake Canyons

was built at the site. The mill only operated for a year before it was dismantled and moved to White Pine. The post office closed on March 13, 1881. Soon after the office closed, the town was abandoned. Total production for the Hot Creek district from 1867 to 1881 was valued at just over $1 million.

The district remained quiet until early 1897, when new ore discoveries prompted some people to return. The post office reopened on May 5, 1897, and continued to operate until January 26, 1912. The Hot Creek Syndicate Trust took control of most of the area's mines. It had thirty-two claims in nearby Rattlesnake Canyon. In 1910 Hot Creek's population was still twenty-five, but after the Syndicate folded in 1911 the town rejoined the ghosts. Not much mining has taken place there since, although Harry McNamara, George Dugan, and John Pinola produced some antimony in the teens. The Hot Creek Consolidated Mines Company, which began operations in the early 1920s, tried to make a profit by reworking some of the old mines, but it was unsuccessful, and in December 1925 its properties were attached with debts of $65,000.

The Hot Creek Ranch Company organized a large ranch encompassing the old Hot Creek townsite. The ranch has been the salvation of the town's remaining buildings. The ranch owners have taken care of the buildings and still use some of them. One of the most impressive structures is a stone building that served as a hotel. Senator Joseph Williams built it in 1908. He married Sophia Ernst in 1870, and they spent the rest of their lives at Hot Creek. Williams also served six years in the Nevada assembly and senate. He died in 1910 and was buried in the family plot of the Hot Creek Cemetery. Sophie died in 1927, and long-time family friend Elizabeth Barndt bought the ranch for $25,000. The remains of the charcoal kilns north of Hot Creek are well worth the trip. They are unique and are some of the few such kilns left in Nye County. A number of other ruins are located higher up in Hot Creek Canyon. For the most part, these are only old homesteads that were not connected with the mining activity.

Keystone
(Florence) (Rattlesnake)

DIRECTIONS: From Warm Springs, head north on US 6 for 8 miles. Exit left and follow the road for 3½ miles. Bear right at the fork and follow the road for 6½ miles. Exit left and follow this road for 2 miles to Keystone.

Keystone was a small but active camp dating from 1868. It served as the company town of the Old Dominion Mining Company. The camp's population soon reached fifty, and residents supported two saloons. However, the initial excitement quickly died down, and for the next twenty years the Keystone mine operated only intermittently.

In February 1881 the nearby Florence, or Rattlesnake, district formed, and

Ruin at Keystone. (Shawn Hall collection, Nevada State Museum)

Preserving the Glory Days

by summer miners were working the Louisa, the Buccaneer, the Florence, the Sedan, and the Oliver Twist mines. Many of the prospectors lived in the old camp of Keystone. Interest in the new district slowly faded, and by 1885 its mines had been abandoned. The Keystone mine was the mainstay of the town. A fairly large mill was built adjacent to it in the late nineteenth century, but little information exists as to its ownership and construction dates. The mill was used heavily from 1910 to 1926 to treat ore from a variety of local mines. During this period about twenty-five people lived in the town.

In December 1910 J.H. Hillyer made a rich silver strike at the 600-foot level of the Keystone mine. Ore from the strike assayed at 1,298 ounces per ton. The Hot Creek Development Company began working the Dexter mine and some other claims. By 1912 the Hot Creek Consolidated Mines Company was also actively working the Dominion mine, and Hillyer had sold out to E.A. Edwards. A post office opened at Keystone on January 26, 1912, and stayed open until March 12, 1927. Worth Wiswall was postmaster throughout this entire period. After the post office closed, ghosts quickly overtook Keystone, although mining still took place there.

Beginning in late 1926 the MacNamara Mining and Milling Company took over Victor Barndt's mines (the Krotons, the Buckeye, the San Francisco, the Detroit, and the California). The Treadwell-Yukon Mining Company, which had a huge operation at Tybo, was shipping ore from a number of mines in the canyon to its mill. But this bout of activity, which lasted into the 1930s, had little effect on the empty town of Keystone. In 1930, however, the Tybo Dominion Mines Company refurbished the old Keystone mill and used it to process its ore for a couple of years. After the mill closed it was dismantled, and only an empty shell was left. During the late 1970s and early 1980s a cyanide leaching operation reworked the old tailings, bringing some new life to the ghostly town. In 1990 Nevada Goldfields, Inc., employed forty-two men in its leaching operation, which has since ceased. A number of old stone buildings are left at Keystone, but the ruins of the old mill were flattened to make things easier for the leaching crews. The site is still well worth the trip, and fresh spring water is available there.

Lauville

DIRECTIONS: *From Antelope (Clear Creek Ranch), head south for 3 miles. Exit right onto a poor dirt road that runs parallel to Wattles Creek and follow it for 2½ miles to Lauville.*

Lauville was a small and extremely short-lived mining camp that was active during the early twentieth century. Like the nearby towns of Antelope and Stargo, it never grew much. Its population peaked at around fifteen. A few stone cabins were built to accompany a small number of flimsy tents. Lauville disappeared from all maps by 1912, and no activity has been recorded there since. Nothing much remains at Lauville except for a few piles of stones left from the cabins. Fresh water is available from nearby springs.

Lockes
(Ostorside) (Kaiser) (Keyser Springs)

DIRECTIONS: *Lockes is located 21½ miles south of Currant on US 6.*

Lockes, first known as Kaiser or Keyser Springs, was a ranching settlement that later developed into a service station and food store. The springs were first used in the 1860s by freight wagons heading from Hamilton to Reveille. In 1875 W. H. Reynolds homesteaded the springs. He sold them to Eugene and Elisha Locke in 1883. Elisha moved to Eureka in 1890, but Eugene remained, and in 1893 he married Sara Ernst, who came from a prominent Nye Country family. In the 1920s they built a gas station and restaurant. Eugene died in 1926. Madison and Charlene Locke later took over the complex. The station and restaurant closed in the 1950s, and the Lockes sold out in 1963. All the members of the Locke family are buried in a nearby cemetery.

A school opened at Lockes in 1935 and was in operation for many years because the settlement was a central location for local ranches. The site was the only place to get gas and food between Currant and Warm Springs. Now the beautiful, green, serene site, which has been abandoned for years, is slowly eroding away. A number of buildings remain at Lockes, and the site is worth a quick stop.

Lower Town

DIRECTIONS: *Lower Town is located in lower Six-Mile Canyon, 5 miles northeast of Hot Creek.*

Lower Town was a very small camp that formed in 1867 after a five-stamp mill was built there. The mill shut down after only a few months, and residents quickly abandoned the camp. Only faint foundations of the mill mark the site. No roads go all the way to Lower Town, and anyone wishing to see its remains will need to hike a long way.

Moore's Station

DIRECTIONS: *From Warm Springs, head north on US 6 for 24 miles. Exit left and follow this road for 15 miles to Moore's Station.*

Moore's Station was a stage stop on the Belmont-Tybo-Eureka stage run. The Shoshone Indians had visited the site long before the stage line began operating. They called the springs at the site *dzicava*, meaning "dried juniper water." The station was established in the early 1870s, soon after the four Moore brothers settled there and established a small ranch. H. A. Moore, a former resident of Tybo who also owned the Twin Springs Ranch, ran the station. His brother William served as the stage line agent, E. C. was stage driver, and Walter was the mail carrier. The men dug a small reservoir, brought in a number of fruit trees, and established one of Nevada's first orchards. For years they were also active in mining interests at nearby Morey, and the station served as a shipping point for goods headed to that camp. After the stage stopped running, Moore's Station no longer had much of a purpose, and the Moores left. In 1898 the *Belmont Courier* reported that the station, then owned by M. S. Sharp, had burned. While the report turned

out to be false, the paper accurately noted that the station had been vacant for years.

The station house and ranch were used as recently as the early 1970s. The author's many attempts to acquire the property have failed. The ranch is currently considered part of the Upper Hot Creek Ranch allotment. It was at one time one of the many ranches that George Russell owned, but it was recently put up for sale as part of the huge Hot Creek Ranch, leaving the future of Moore's Station in doubt, because it is not clear that whoever purchases the ranch will respect the building.

The remains of Moore's Station are extremely impressive. The stage house is still in excellent condition, although that could change, since no one is presently living there. A number of sod-type buildings and a few stone buildings and remains are left at the site. Two stone buildings, one that was used for storage and one that was used for a garage, are both in excellent condition. A few small stone dugouts also remain, along with the sod ranch buildings. The orchard is in poor condition but continues to bear fruit every year. Running water is still available at the site. Moore's Station has been one of the author's favorite ghost towns since he first visited it in 1979.

Morey

DIRECTIONS: From Warm Springs, head north on US 6 for 20.9 miles. Exit left and follow the road for 3 miles. Take the right fork and follow the road for 7¾ miles. At the intersection, continue going straight for 3½ miles. Go left at the fork and follow the road for 3 miles to Morey.

In 1865 a large group of prospectors from Austin came to southeastern Nevada. T.J. Barnes, a member of this group, discovered rich ore only six miles from Moore's Station. S.A. Curtis, John Emerson, and William Muncey organized the Morey Mining District the following year. Although the *Reese River Reveille* printed a favorable report of the ore find in 1867, the Morey camp did not actually form until 1869. In 1867 the American Eagle, the Magnolia, the Eureka, and the Mount Airy mines, all of which the Hall and Emmerson Company owned, were producing. In early 1869 the men who had organized the district sold a four-fifths interest in their property to an eastern company for $35,000. The buyers formed the Morey Mining Company. Ore was shipped to the Old Dominion mill in Carrolton. The district gradually grew, and on November 15, 1872, a post office, with David Ogden as postmaster, opened in Morey.

In 1874 the Morey Mining Company completed a ten-stamp mill, but it only functioned for two months before closing down. Construction of the mill was actually completed in February, but it didn't begin operating until October because of litigation over payment for services. It was forced to close in December when it couldn't pay its employees. The remaining bullion bars were confiscated. Ore was then sent to mills in Tybo and Belmont. In 1874 Morey's population reached ninety-five. In addition to its own post office, Morey also contained two stores, a blacksmith shop, a livery stable, an express office, and a boardinghouse run by Charles Dexter. A daily stage line ran from Morey to Belmont and Eureka. Supplies for the camp were brought in from Eureka at a cost of $35 per ton. However, by the end of 1875 most people had left Morey, and there was little additional activity in the area until 1880.

A new company, the Morey District Mining Company, took over the mines. Max Bernheimer was superintendent. The mill reopened from April to December 1880, producing $81,000 worth of gold and silver. The biggest problem was that the ore was a heavy antimonyl sulphide, which meant that it had to be burnt in pits before being treated in the mill. Along with thirty-five other claims, there were seven major mines in the district. They were the Little Giant, the Black Diamond, the Monterey, the American Eagle, the Bay State, the Cedar, and the Kaiser. The deepest shaft was the American Eagle, which reached 370 feet. At 1,100 feet, the longest tunnel was the Bay State. The ore from these mines was sent to Eureka by freight wagon once the mill had closed for good. In the 1880s Morey's population remained at about thirty, and various owners worked its mines. In 1889 the Airshaft, the Point of Rocks, the Magnolia, and the Morey mines produced enough to warrant the restarting of the old mill in the summer. In response to demand, a school opened in the camp. At this time, H.A. Cohen ran most of the mines, but he had to cut his work force because of the depressed price of silver. Finally, in 1894, he gave up and relocated to Pioche. In 1899 Ernest Schandel returned to Morey after a ten-year absence and reopened the mines, but he too gave up in 1905.

After the post office closed on April 15, 1905, the district was almost entirely abandoned. By 1909 Morey was empty. The new Morey Mining Company, owned by George and J.W. Wist from Manhattan, staged a small revival in 1919, working a new property, the Smuggler, and an old one, the Airshaft. In addition, R.C. Gaston and John McGiluroy relocated some of the other old mines, including the Silver Dyke, in July 1922. In January 1925 Dr. G.F. Tilson bought the Morey mine and fourteen other claims for $150,000 but had little success. The revival lasted a few years, but by 1926 Morey was a ghost town again. In 1934 Victor Barndt and E.M. and Lee Booth took over the Morey mines and made one last attempt to find additional rich ore, but their efforts were fruitless. The mines are still tested once in a while, but no sustained activity has occurred. Total production value for the Morey mining district stands at a little over $475,000. Only ruins of stone cabins and some wood mine hoists mark the site of almost-vanished Morey. Some remains of the town were wiped out by a huge landslide that occurred during the 1950s when an underground nuclear test was conducted on the flat area below Moore's Station. The shaking triggered some previously unknown faults and caused a section of Morey Mountain to let go. Evidence of the slide is visible today.

Nyala
(Mormon Well) (Polygamy Well) (Sharp Ranch)

DIRECTIONS: *From Warm Springs, head east on Nevada 375 for 15.9 miles. Exit left onto the Currant–Nyala road and follow it for 22.1 miles to Nyala (Sharp Ranch).*

Nyala was a small ranching settlement that also served as a rest stop for weary travelers. In the 1860s a well at the site provided water for horse teams traveling through hot, dry Railroad Valley. The well was known both as Mormon Well and as Polygamy Well, and it was used as a stop along the Midland Trail during the "Great Race" of 1908. In 1913 a regular settlement began to form at the well site when Herman and Alvena Reischke moved there and set up a small ranch as well as a boardinghouse, a general store, and a restaurant. A post office, with Herman Reischke as postmaster, opened on February 5, 1914, and the settlement was renamed Nyala. Alvena Reischke suggested the name Nyala after he saw a photograph of the East African antelope that bore the name. He thought the animal so beautiful that he wanted their little settlement named after it. The Reischkes continued to work the Nyala Ranch complex until August 1917, when they sold out to men named Goodman and Crosby. The new owners ran the ranch until the 1920s. That year, George Sharp, who had been buying most of the ranches in Railroad Valley, purchased the Nyala complex. Sharp and his wife, Mary, kept the ranch in operation but closed the store and restaurant.

George Sharp died in 1933 during a trip into the desert, and his eldest son inherited the Nyala Ranch. The post office closed on January 15, 1936, but the ranch remained a consistent cattle producer. The ranch is still in the Sharp family. Most of the original buildings, along with some newer ones, are in use today.

All residents of Nyala in 1914. The tents to the left contained a mercantile store and post office. (Nevada Browne collection, Central Nevada Historical Society)

The still impressive Pritchard's Station is in the middle of nowhere and is a pleasant reward after the short hike required to reach it. (Shawn Hall collection, Nevada State Museum)

Pritchard's Station

DIRECTIONS: *From Moore's Station, continue north for 1 mile. Bear left and continue for 4¾ miles. Exit right and follow the road for 1½ miles to Pritchard's Station.*

Visitors to Pritchard's Station will be richly rewarded for traversing a soft, treacherous road that requires a four-wheel-drive vehicle. Sometimes even a four-wheel drive can't navigate it. The old stage station is in excellent condition. It was located on the old Belmont-Tybo-Eureka stage line and was active from the 1870s to the 1880s. On February 5, 1874, a post office, with Kate Shoop as postmaster, opened at Pritchard's, but it closed on April 14. In addition to the station, there are a few old corrals and an old foundation, which are apparently the remains of a stable. The site is a definite must, and is well worth the hair-raising drive to get there.

Sawmill

(Willows Creek Ranch) (The Willows)

DIRECTIONS: From Moore's Station, continue north for 5 miles. Follow the left fork, keeping right, for 24 miles to Willow Creek Ranch, the site of Sawmill.

As one might assume, the small camp of Sawmill grew up around a sawmill. During their booms years in the 1870s the mines at Morey, Tybo, and Hot Creek were the main users of the wood the mill processed. The operation never really expanded, and later the site was incorporated into a ranch. In the 1870s Sawmill also served as a horse-changing stop on the Eureka-Tybo-Belmont stage line. A stage robbery was attempted there in September 1877. One robber was killed, and two escaped. Shotgun blasts injured the Wells Fargo agent, but he survived. The Willow Creek Ranch, which is smaller than the original one, has been operating intermittently over the last twenty-five years. A number of ranch buildings, including the old ranch house, remain at the site. There is no trace of the sawmill.

Silverton

(Black Rock Summit)

DIRECTIONS: Silverton is located 8 miles south of Lockes on US 6.

Silverton was a short-lived mining camp that sprang up in 1921. Joe Tognoni actually discovered the Silverton mine in 1914, but little development occurred. In May 1921 Harry Stimler and Archie Connor made a new strike on Tognoni's claims, and the Silver Hills Extension Mining Company bought the property. A small tent camp quickly formed on a little mesa below the mines. By June some frame buildings were already replacing the tents, and Pete Beko opened a store. The Treasure Hill Mining Company, financed by Silver Hills, formed to work the newly discovered Treasure Hill mine, which Stimler had also located. The company bought equipment for the district's mines from the Gibraltar Mining Company and hauled it to Silverton from Jett. However, by the spring of 1922 the ore pockets had completely disappeared, and operations stopped. Only Tognoni remained in Silverton, making occasional small shipments. Tognoni died in 1932 and was buried at Silverton. No other activity has taken place there since. The limited mining that occurred there produced ore worth only $20,000. Silverton's buildings were relocated, and today only foundations and mine dumps mark the site.

Stargo

DIRECTIONS: *From Antelope, continue north, bearing right, and follow the road for 4 miles. Exit left and follow this poor road for 6 miles to Stargo.*

Stargo was a little-known mining camp that came into being on Stargo Creek in Little Fish Lake Valley. The mining camp formed in 1908 after prospectors discovered gold, silver, and lead ore in December 1907. Miners organized the Stargo Mining District in January, and fifty claims had been filed by the end of the month. But deposits of rich ore large enough to sustain the camp weren't found, and Stargo died as quickly as it had arisen. A few small stone and wood cabins were built during the site's short period of activity. The camp had vanished from all maps by 1910, and hardly a trace of it remains today. Only one completely flattened stone cabin marks the site.

Summit Station

DIRECTIONS: *Summit Station is located 10 miles north of Pritchard's Station.*

Summit Station was an original stop on the Belmont-Tybo-Eureka stage line. It remained in existence only a short time, because in the early 1870s, a few years after the station was built, the route changed, and Hick's Station replaced it. Summit Station was then abandoned. The site was never used again, and only the ruins of the stone cabin that served as the station are left.

Sunnyside
(Whipple Ranch)

DIRECTIONS: *From Lund (in White Pine County), head south on Nevada 318 for 30 miles to Sunnyside.*

Sunnyside was, and is, a small settlement in the southern part of the White River Valley. It started as a ranch owned by a man named Horton. He built the ranch in the 1880s, and soon a small settlement surrounded it. A post office, with Mary Horton as postmaster, opened on July 10, 1880, and served the whole White River Valley. The small settlement never really grew, but it did have a small mercantile store that provided supplies for the many ranches in the valley. Jewett Adams, who would later become governor, had a ranch nearby in the late 1880s and the 1890s.

John Whipple bought the Horton Ranch in the fall of 1904. He moved his family to Sunnyside from Lund after Christmas of that year. Whipple had 200 head of beef cattle on the ranch in 1904 and 1905. He sold them in DeLamar, a mining town in Lincoln County. Whipple took over O. H. Snow's mail contract in 1904 and ran the route from Pioche to Sunnyside, a trip that took two days by buggy.

Whipple worked this route for two years before hiring a man to do the job. By a quirk of fate, this decision turned out to be very profitable for Whipple. The hired man lost a horse on one of his trips in February 1906, and Whipple had to go and search for the animal. While searching, he found an outcropping of rock that turned out to be high-grade silver ore. This was later developed into the Silver Horn mine, four miles north of the mining camp of Bristol. Whipple sold his mine to an English mining company in March 1906 for $10,000 and used the money to pay off the mortgage on the ranch. He maintained an interest in the mine, which provided him with a small cash flow. Whipple also had a group of claims, the Silver Dale group, located next to the Silver Horn mine. He sold these claims in late 1906 to the same English company for $3,500.

Whipple was officially appointed postmaster at Sunnyside on September 27, 1917. He held the post until January 31, 1933, when the office closed. For many years, Sunnyside was the site of a school. When the ranch was short of hands, the teacher would cancel classes and help out.

Whipple bought back the Silver Horn mine and the Silver Dale claims in the spring of 1920 but spent most of his time at Sunnyside. His son, Clair, leased the ranch from Whipple in 1935 and purchased it in 1940. In 1945 Sunnyside still had a population of twenty-seven. Most of the residents worked on the ranch. Clair Whipple and his wife, Lila, ran the ranch until 1962, when they sold the property to the Johnson family, the present owners. Not much remains at the old Sunnyside ranch except for a few log ruins. The ranch is in a beautiful setting and is well worth the trip. Camping facilities are available at the Warm Springs Camp, southwest of Sunnyside. Sunnyside is also the base for the Wayne E. Kirch Wildlife Management Area. Tremendous wildlife viewing opportunities exist throughout the area.

Troy

DIRECTIONS: *From Nyala, continue north on the Currant–Nyala road for 9 miles. Exit right and follow the road for 3 miles to Troy.*

Alexander Beaty discovered the first ore in the Troy area in May 1867. He organized the Troy Mining District on October 27, 1868. William Murray was elected district recorder. Beaty sold his claims to a Yorkshire,

England, mining company in 1869 and moved to Butterfield Springs, where he started a prosperous ranch. The Old English Gold Corporation built a twenty-stamp mill in Troy in 1870 and began operations. Harry Newton built the stamp mill five and a half miles west of the Troy mine. It cost more than $500,000 and was equipped with Stetefeldt furnaces. It was finally completed in 1871.

The camp that grew around the huge stamp mill had a few stores, an express office, a boardinghouse, and some cabins. An unofficial post office formed in 1870, but it was not until February 18, 1873, that an official U.S. post office, with Charles Palmer as postmaster, opened. By 1871 the population of Troy was well over 100.

The English mill handled ore from both the Troy and the Locke mines. The Troy mine comprised two separate mineral lodes, the Troy and the Gray Eagle. The mine shaft was more than 500 feet deep, with many branch tunnels. Although the Troy mine was more than 1,000 feet above the camp, water seepage into the mine was a constant problem and eventually led to its abandonment in 1913. The Locke mine, three quarters of a mile south of the Troy, had a 500-foot tunnel and no water problem. Unfortunately, however, the mine was never a big producer. The two other main mines in the district were the Clifton and Blue Eagle. The Blue Eagle had a 500-foot shaft and a 700-foot branch tunnel. Overall, though, the district's only big producer was the Troy mine. The ore mined in the area contained silver, gold, copper, and small amounts of lead.

Not much is left of the once impressive Troy Mill. (Shawn Hall collection, Nevada State Museum)

Preserving the Glory Days

The stamp mill operated for six months until it closed in 1872 because of extremely low-grade ore. When the Martin White company bought the mill for $11,000 in 1876, it sent the equipment to Ward in White Pine County. The English company pulled out of Troy soon afterwards. Around seventy miners stayed on to prospect the different sites along the canyon. The post office operated until August 7, 1876. Even after it closed, some people still remained, although no ore was shipped out. Freight arrived once a week from Eureka, indicating that faint interest in Troy still existed. A small-scale revival started in 1908 when prospectors found new ore deposits in the Locke and the Blue Eagle mines. The post office, with John Evans as postmaster, reopened on April 24, 1908, and managed to stay open until February 28, 1913. The amount of ore taken out of the mines was small, but it was enough to keep the town going a little longer. The revival slowly died out, and by 1915 Troy belonged to the ghosts. The town woke from its sleep for a short spell when the call went out for metal prior to World War II. Joseph Hafen, who organized the Troy Silver Mining Company, reopened the mines in 1936 to help boost prewar production. He built a twenty-ton flotation mill in 1938, and it was enlarged to fifty-ton capacity ten years later. The company maintained fairly steady production until 1949, when it shut down. A rise in mineral prices brought the company out of mothballs the next year, and minor production resumed. In 1956 miners discovered uranium in the gold vein the company was working, but the gold and uranium vein disappeared soon afterwards, and the company folded for good.

Troy still has a few inhabitants. Three of the remaining buildings are occasionally occupied by hunters and fishing enthusiasts. The stream flowing through the ruins and down the canyon is full of fish. There are five standing buildings, four of which are made of stone and date back to the original settlement, and the other which is a remnant of later operations. The ruins of the mill, directly opposite the buildings, are comprised of the smelter stack, huge concrete foundations, and the unique remains of the Stetefeldt furnaces. Many faint foundations, as well as a number of interesting dumps, remain at Troy. The Locke mine still has all its mining apparatus. Most of the mine workings are left from recent operations. The road to Troy is fairly rough but is passable with a four-wheel-drive vehicle. The first sign of Troy is the Locke mine, located high up on Troy Mountain.

Twin Springs

DIRECTIONS: *From Warm Springs, head east on Nevada 375 for 10½ miles to Twin Springs.*

Twin Springs is a small ranching settlement that dates from the late 1870s. The ranch at Twin Springs is of historical interest because it has been active for more than 100 years. H.A. Moore, who also owned Moore's Station, ran the Twin Springs Ranch until the 1890s. The Fallini family has owned the ranch for a few generations. A number of old buildings still remain at the ranch, along with newer ones. The present population of the ranch is about ten. A school still operates periodically at the ranch. Beware: sudden water fights have been known to break out at a moment's notice when people are visiting! The site is interesting because it is still set up much as it was long ago.

Tybo

DIRECTIONS: *From Warm Springs, head north on US 6 for 8 miles. Exit left and follow the road for 3½ miles. Take the left fork and follow the road for 3 miles to Tybo.*

Prospectors first made discoveries in the Hot Creek range in 1866. One of the discovery sites was named Tybo, a word derived from the Shoshone *tai-vu,* meaning "white man's district." Prospectors made another discovery nine miles south of Tybo, and the Milk Springs Mining District was organized in February 1867. Miners then worked the Bismark and the Fisher mines for a short time before abandoning them. Dr. Galley and M.V. Gillett made the first major ore discovery in Tybo in 1870. The discovery was later developed into the Two-G mine.

It was not until 1874 that a small camp began to form in Tybo Canyon. The first settler in the canyon was John Centers, who moved there in August 1866. A small lead smelter was built in the canyon in 1874, and the Tybo Consolidated Mining Company formed the following year. The company built another smelter and a twenty-stamp mill before the end of 1875. The mill was the old Highland mill from Pioche. Tybo Consolidated controlled three major mines near the camp; they were the Casket, the Lafayette, and the original Two-G. The Two-G was the deepest of the three, having been dug to a depth of 450 feet before closing in 1883. In 1875 T.J. Bell began running a twice-a-week stage to Eureka. A Wells Fargo office was established in the growing camp. Businesses in town included the Trowbridge Store, the Rosenthal Store, Barney McCann's Restaurant, and the W.F. Mills and Company Bank. Trowbridge was the biggest businessman in Tybo. In fact, in 1879

the owners of the Two-G mine grew so indebted to him that he became the mine's principal owner.

By the summer of 1876 the small town's population had grown to almost 1,000. Because of constant problems arising from the clashing of cultures, the town had been divided into three separate sections: an Irish section, a Central European section, and a Cornish section. Buildings in the town in 1876 included five stores, a number of saloons, two blacksmith shops, and a post office—with Charles Barrett as postmaster—which opened on September 3, 1874. The *Tybo Sun*, under the management of a Mr. Ragsdale, began publication on May 19, 1877. Ragsdale sold the paper in early 1878 to William Taylor, who operated it for a short time and then sold out to William Love and D. M. Brannan. They continued to publish the weekly until declining interest forced it to fold on March 14, 1880. Tybo was basically a company town of the Tybo Consolidated Company, a company that had a great reputation with its workers. In February 1877 the employees presented superintendent Matt Howell with a gold watch as a token of their appreciation. Activity in the town continued to increase in 1877. A new brick schoolhouse was completed in April, a jail was built in the spring, Bell's stage began running three times a week, the Tybo Brass Band formed, and an International Order of Odd Fellows (I.O.O.F.) chapter organized. By 1877 businesses in Tybo included the William Tell Saloon, owned by J. Morasci; the Pioneer Market, owned by Stonebarger and Frakes; the Court Saloon, owned by John Wheatly; the Exchange Saloon, owned by the Gilmore Brothers; the Tybo Bakery, owned by J. Gilman; the Tybo Brewery and Saloon, owned by

A couple of the numerous surviving charcoal kilns in the hills around Tybo. (James O'Neill collection, Central Nevada Historical Society)

Valentine Lechner; the N.J. Devine Blacksmith Shop; the Erie Lodging House, owned by Mrs. S. Hawes; the Rosenthal Store; the Delmonico Restaurant, owned by Catherine Williams; the Tybo Drug Store, owned by J.S. Hammond; the Palace Saloon, owned by Kind and Everett; the Peoples Saloon, owned by John Peoples; the Pacific Hotel, owned by F.O. Swensson; the W.H. Clark General Merchandise; Luse's Restaurant; the 2-G Boardinghouse, owned by R.N. Oliver; the Tybo Restaurant, owned by James McFadden; the City Drugstore, owned by Garrett and Joslyn; and the Headquarters Saloon, owned by Ferguson and Reilly. The Tybo Literary Society was active from 1876 to 1881. One of its members, Mary Godat, later became the first female legislator from Wyoming.

In May, Henry Allen built fifteen kilns up in the canyon. It took more than 500,000 bricks to complete construction. The new kilns replaced the stone ones that had been built in 1874. In September the Tybo Company drastically cut back its operation, and many people left town over the winter. Early in 1878, however, new discoveries in the old mines made it necessary to hire a bigger work force. The mines produced ore worth more than $100,000 in February 1878 alone. The town's weekend hot spot was Spence and Brougher's Coal Burners Saloon.

In 1879 the Tybo Consolidated Mining Company decided to close its two smelters, which employed 400 people, and build a crushing and roasting mill. The mill had a daily capacity of 80 tons but did not employ nearly as many men. This caused a drop in population, but three hotels, two restaurants, an express office, and an assay office were built in 1880 nevertheless. From 1877 to 1880 Tybo was Nye County's top producer and was second only to Eureka in total lead production for the entire state.

The Tybo Consolidated Mining Company ran into problems in early 1881 when the quality of the ore dropped drastically. The mill closed down, and the company only worked the mines, but it was forced to fold in November. It sold its equipment, which was moved to Bristol (in Lincoln County). By the end of 1881 only 100 people were left in Tybo Canyon, and the future looked bleak. During the next twenty-five years, Tybo barely managed to cling to life. A number of different mining enterprises attempted to profitably work the Tybo district, but all of them quickly failed. In the 1880s and 1890s most of the activity in Tybo was limited to small groups of residents working some of the 100 claims made during the town's boom. A major fire struck the half-empty town in July 1884, destroying thirty-two buildings for a loss of $33,500. During that year the mines only produced $13,000 worth of ore. In the 1880s and 1890s the mill was still used sporadically. In 1887 the

View of the Tybo Consolidated Mining Company's smelter and 20-stamp mill in 1875. (Central Nevada Historical Society)

Nye Mining Company gained control of the mines and hired a work force of thirty-five people. N.S. Trowbridge and Joseph English, both long-time residents of and businessmen in Tybo, owned the company. But their success was limited, and the company folded in May 1889. In 1890 the Diminick and Ma Alta mines reopened, and the mill started up again, but all work stopped in January 1891 when the company couldn't pay its employees. By 1894 there were only sixteen people left in the town. The post office remained open through all of this and continued to serve the dwindling population until July 14, 1906.

When the Nevada Smelting and Mines Corporation started work in May 1906 with $5 million in capital, the town was finally rewarded for its perseverance, if only for a while. Max Berrheimer was president of the company, which took over the old Tybo Consolidated Mining Company's holdings and began new operations. But profits were extremely small, and the company left the district in 1908, leaving the town almost completely abandoned. There were only four residents remaining in Tybo by 1911.

Tybo received another lease on life when the Louisiana Consolidated Mining Company began to work a few of its mines in 1916. The company, which was incorporated in May 1912 and of which Julius Sieghart was president, owned property in Tybo as well as in Reveille and in the Oneota district (in Mineral County). The company obtained a twenty-year operating agreement on the Tybo mine and the nearby Diminick mine. Soon a fleet of ten trucks was hauling ore to mills in Tonopah. Water was a problem, and the company was pumping 600,000 gallons a day. The company employed more than forty men. A 100-ton concentration mill, built in 1917, operated until late 1918. At the end of 1919 the company installed a flotation plant and a lead smelter. These operated until 1921. The company brought electricity, telegraph, and telephone service to the town in March 1920 via a line strung from Millers, in Esmeralda County, at a cost of $100,000. By July 1920 the company had a workforce of eighty. The company faltered in 1921, and by 1922 the mines had stopped producing ore. The company left the district in 1922 to work its other properties. Soon Tybo was empty again, and only rows of deserted, decaying buildings attested to the fact that people had ever been here. In December 1924 the Manhattan-Tybo Power Company discontinued service to Tybo.

Tybo's last revival started in 1926, when the Keystone-Hot Creek Mining Company purchased property in Tybo Canyon. The company leased this property to the Treadwell-Yukon Company, Ltd., which built a 350-ton concentration mill and a new smelter in 1929. The mill was soon processing more than 300 tons of lead ore daily. Another company, the Tybo Dominion Mines, Inc., also became active in the district, purchasing the Old Dominion and the Jumping Jacks claim groups. The company was incorporated in 1928, and

Irving Farrington was president. Operations never really got off the ground, however, and the company quietly folded in 1929.

The Treadwell-Yukon Company became the core of the Tybo revival. More than seventy-five people came back to Tybo in 1929, and the post office, with Willard Hales as postmaster, reopened on February 11. Treadwell-Yukon built a number of two-story boardinghouses for the miners and brought in some prefabricated houses for the executives. There were 180 men on the payroll by the end of 1929, and Tybo's population had grown to 228 by 1930. Over the next eight years the mill processed more than 500,000 tons of lead ore. The company sank a 1,500-foot shaft during that period. While digging it, miners discovered some small silver deposits. The revival ended in 1937 when the company closed the mill. It was dismantled shortly afterwards, and the United States Machinery Company purchased all the equipment. From 1929 to 1937 Tybo's mines produced $6.8 million worth of ore. From 1942 to 1945 eighteen men worked hauling old tailings to Tonopah for treatment. That was the last thing the Tybo district produced. Tybo's total production value stands at an amazing $9.8 million—amazing because during Tybo's best years of production, lead and silver prices were extremely low. If the ore were being mined today, its value would be almost triple what it had been at the time.

Tybo is one of the better ghost towns in Nye County. A handful of people still make their homes in the peaceful, beautiful canyon, throughout which extensive mine and mill ruins are scattered. Many buildings remain, and most are still in very good condition. Among the better remains is the old Trowbridge store, which became a miners' recreation hall during the town's last revival. The cemetery is on a small hill at the mouth of the canyon and should not be missed. The Tybo charcoal kilns, which are further up Tybo Canyon and are difficult to reach, are well worth the tough drive.

Warm Springs

DIRECTIONS: Warm Springs is located on US 6, 49 miles east of Tonopah.

Warm Springs was originally a stopping place for freighters and stages traveling to Eureka and Elko. The first settler came to the site in 1866 and built a small stone house next to the warm, soothing springs. The small settlement never grew much, but after the turn of the century a store and lodging house were built there. Warm Springs continued to serve a small number of weary desert travelers. During the 1920s the town reached its peak. On January 19, 1924, a post office, with Ethel Allred as postmaster, opened at Warm Springs. Although it closed on June 29, 1929, there always was, and still is, a fairly constant number of travelers stopping at Warm Springs.

Nobody lives at Warm Springs today. A small trailer park that existed in the 1970s is now gone. The combination saloon and gas station has also

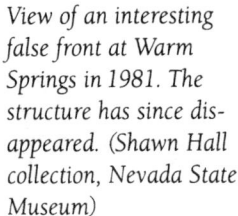

View of an interesting false front at Warm Springs in 1981. The structure has since disappeared. (Shawn Hall collection, Nevada State Museum)

Preserving the Glory Days

closed. Once the leaching operations at Reveille and Keystone stopped, the small amount of traffic on the highway couldn't sustain the operations. A number of old buildings from Warm Springs's early days still remain. The springs still flow, and while a private swimming pool is off-limits to visitors, a couple of small huts are set up in which the traveler can enjoy a warm dip in solitude.

Willow Creek

DIRECTIONS: From Nyala, head south on the Currant–Nyala Road for ½ mile. Exit left and follow the road for 2¼ miles to Willow Creek.

The Willow Creek Mining District is in the Quinn Canyon Mountains. Willow Creek, although it is barely remembered today, was once the scene of quite a bit of activity. Charles Sampson and David Jenkins located the first mining claim there in June 1911. They named their claim the Rustler and worked it until the middle of 1912. During this period miners were working many other claims, and by the beginning of 1913 a camp of fifty had come into being. In January the Willow Creek Mining Company, of which Zeb Kendall was president, formed, and it bought the Jenkins and Sampson claims for $25,000. By February Willow Creek had a population of eighty. A discovery of free gold in March 1913 created a great deal of excitement in Warm Springs. Steve Pappas and W. Blackwell came upon it in the Melbourn vein, located at Gold Springs. In April George Wingfield bonded the gold claims of W.C. McMullen for $100,000. In June Wingfield bought the Willow Creek Mining Company, which had already shipped $20,000 worth of ore, and he quickly had two shifts working. In September the company shipped 1,000 sacks of ore, and the Willow Creek Gold Mining and Milling Company, of which John Eugelke was president, organized to work the Blackwell and Pappas claims. The following month the company was rewarded with a rich vein that assayed as high as $78,000 a ton. The first 500 pounds of ore returned $2,960, but because of winter weather, nothing else was shipped until spring.

The Willow Creek district comprised seven separate claim groups. The first was the Gold Spring claims, located in Gold Canyon, two miles north of Osterlund Camp in Willow Creek Canyon. The Gold Spring group was made up of five claims that had vertical veins containing free gold laced with large amounts of green talc. A twenty-five-ton mill was built in Gold Canyon in 1914 and produced $10,000 worth of ore before it closed in 1915.

The Last Chance claim was about one-eighth of a mile west of the Gold Spring group. This claim was the property of William Wittenburg and Dr. Weller, both residents of Goldfield. Miners dug a crosscut tunnel to hit a six-inch vein of gold-bearing white quartz. The two men abandoned the claim after the tunnel reached 175 feet and still wasn't looking very promising.

The Willow Creek Mining Company was the Willow Creek district's main developer. George Wingfield formed the company after he purchased the Rustler claim and five other nearby claims. The Rustler became the company's main mine. The company made its biggest shipment in October 1913 when a 500-pound shipment of free gold ore returned almost $20,000. Altogether, the Rustler mine produced just over $50,000 worth of ore.

The remaining four claims—the Queen of the West, the Mayflower, the Melbourn, and the Battle Axe—were smaller workings and, except for the Melbourn, were not very productive. Ore at the Battle Axe claims ran only $6.40 a ton. Activity in the Willow Creek district continued until late 1914, when the Willow Creek Mining Company stopped production. By 1916 only Charles Binimiss, John Steiner, and Frank McMullen were working claims in the district.

The area remained dormant until 1917, when new discoveries of rich gold ore warranted reopening the district. The Gold Spring claims were purchased by the Gold Spring Mining Company, which W. L. McMullen of Nyala managed. In addition to the claims, in November the company started a five-stamp amalgamation and concentration mill that had been moved from Gold Mountain in August. The mill produced concentrate worth as much as $100 a ton. The company worked the Gold Spring claims until 1922 and then folded. The Willow Creek Gold Mining and Milling Company was active from 1917 to 1919 but sold out to Fred Vahrenkemp and Dr. Bowes. From the late teens to 1920 Wingfield also returned to rework the Rustler mine.

The district was quiet until 1926, when the Nyala Gold Mines Company, of which Paul Thorne was president, formed. The company purchased a gold mine adjacent to the Nevada-California Metals Corporation's properties. The gold ore removed from the mine assayed from $200 to $2,000 per ton. The company also purchased the five-stamp mill that the Gold Spring Company had previously owned and started it in June 1927. Nyala Gold Mines planned to build a sixty-ton mill, but the ore faded before construction began, and it dropped the idea. The company folded in 1927, leaving the Nevada-California Metals Corporation as the only company working the Willow Creek district.

The Nevada-California Metals Corporation, of which F. A. Cassaday was president, formed in 1927 and purchased eight claims, including four from Steve Pappas, at Willow Springs. The gold ore, mined from a 600-foot tunnel, assayed as high as $75 a ton. After an initial investment of a few thousand dollars, the company cleared $11,000 before folding in late 1927. Other activity in the district took place in 1928, when the Gaston Gold Company began limited operations on a group of claims south of the Nevada-California claims. The company curtailed its activity in late 1928, and the district was totally abandoned. After World War II the Willow Creek Mining Company formed and employed five men to work the old mines. Operations ceased in 1950 after limited production. The district has been abandoned ever since.

Scattered ruins remain at Willow Creek. Only one structure still stands, but piles of rubble dot the site. The Gold Spring site in Gold Canyon is marked only by small piles of tailings. The road tends to be very rough. Exercise extreme caution.

North Central Nye County

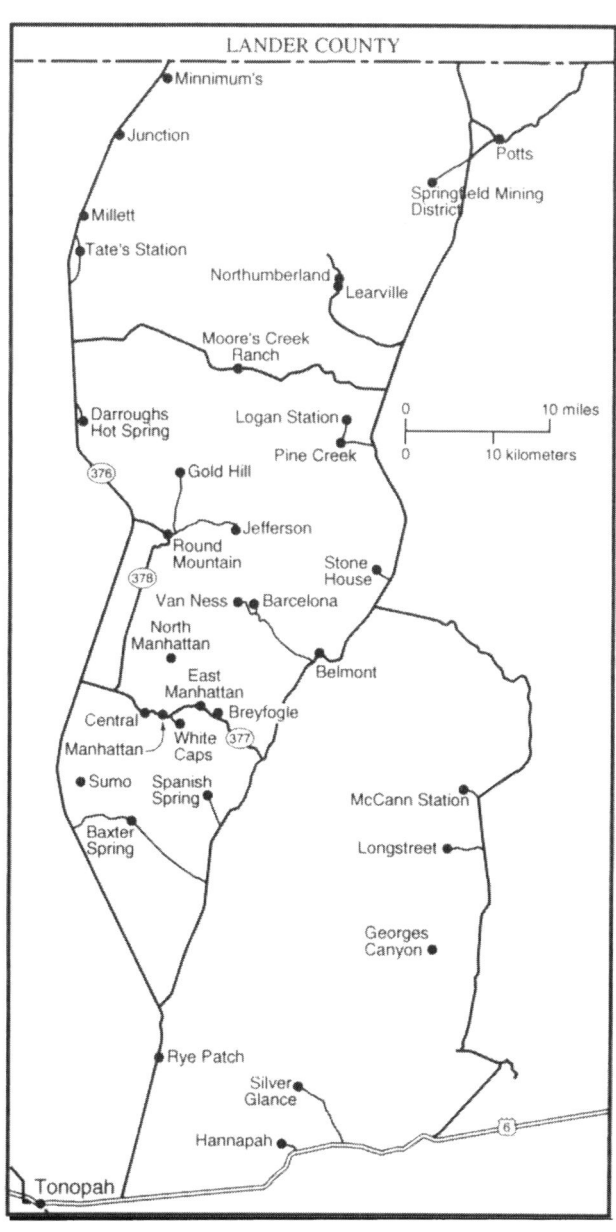

LANDER COUNTY

Minnimum's

Junction

Potts

Springfield Mining District

Millett

Tate's Station

Northumberland ● Learville

Moore's Creek Ranch

Darroughs Hot Spring

Logan Station

(376) Gold Hill

Pine Creek

Jefferson

Round Mountain

Stone House

(378)

Van Ness ● ● Barcelona

North Manhattan

East Manhattan

Belmont

Central

Breyfogle

Manhattan

White Caps

(377)

Sumo

Spanish Spring

McCann Station

Baxter Spring

Longstreet

Georges Canyon ●

Rye Patch

Silver Glance ●

(6)

Hannapah ●

Tonopah ●

0 _____ 10 miles

0 _____ 10 kilometers

Barcelona

DIRECTIONS: *From Manhattan, head east on Nevada 377, bearing left ¼ mile out of town. Follow this road for 7 miles. Exit left on Silver Creek Road and follow it for 8 miles. Remains are scattered for one mile beginning at this point.*

Barcelona, initially a Mexican silver mining camp, first formed in 1874. The Mexicans made discoveries there as early as 1867, but it was not until Castilian grandee Emanuel San Pedro led a Mexican prospecting group into the Spanish Belt district that sincere interest in the Barcelona area arose. In 1874 the Barcelona mine and a number of other smaller mines began operating. A fairly good-sized camp formed and soon had a population of almost 175. Quite a few buildings were erected, including an assay office, three boardinghouses, and a number of the usual business establishments.

Barcelona's boom quickly faltered as the ore faded. San Pedro left and went on to make important discoveries at Grantsville. By 1877 the town had only a handful of residents, and by the following spring it was deserted. In 1880 Barcelona was the scene of renewed activity, but it lasted only a short while, and soon the town was abandoned once again.

There were a number of short bursts of activity in Barcelona throughout the next five decades. The last revival, beginning in 1917, was the most significant and productive. The Spanish Belt Consolidated Silver Mines Company formed that year and opened a number of mines, including the Barcelona and the San Pedro. Jules Barnd, the president of the company, was also president of the Carrara Marble Company. Unfortunately, the superintendent, D. H. Walker, died suddenly at the age of thirty-eight. The company, which employed twenty-five, survived. In July 1918 it hit the main vein at 1,275 feet and received its reward. By the beginning of 1920 the tunnel was 4,200 feet long. Two branch companies, the Spanish Belt Extension Mining Company — of which Warren Richardson was president — and the West Spanish Belt Silver Mining Company, formed in 1920. All told, the company spent $300,000 on developments, including the purchase of a mill and extensive work in the San Pedro and Barcelona mines, which were eventually connected. In addition, the Belmont Big Four Mining Company employed twelve and worked a couple of mines. The Spanish Belt mine was listed on the San Francisco Stock Exchange. In May 1920 an impressive ten-stamp mill was completed at Barcelona, and a concentration and flotation plant was added in 1921. For a while it seemed as if the town was on the revival trail, but after only a year of activity the mill had to close when ore values plummeted. H. A. Stevens bought the company's property for $17,000 at a bankruptcy sale in November 1923. He did little with his purchase and in April 1926 sold the property for $1,700 to Horace Campbell. The mill was dismantled in June

and moved to Clifford. There hasn't been any activity in Barcelona since. All told, the Barcelona mines produced $1.6 million worth of ore.

Until the 1980s, when a small-scale operation reworked the Barcelona mines, the site remained empty except when some leasing activity took place there. The actual town of Barcelona is located about a mile above this operation. A number of old stone and adobe ruins remain at the townsite, along with a few of the old mine shafts. At the Barcelona mill site, other stone foundations remain. There are also a few wooden cabins from the last revival. The site is well worth the trip, and fresh water is available at Barcelona.

Baxter Spring
(Baxter's) (Cedar Spring)

DIRECTIONS: *From Tonopah, head east on US 6 for 5½ miles. Exit left onto Nevada 376 and follow it for 13.3 miles to the junction of old Nevada 82. Follow Nevada 82 for 9 miles. Exit left onto a road marked by Baxter Spring signs and follow it for 6½ miles to Baxter Spring. More ruins are located half a mile past the spring.*

Baxter Spring, also known as Cedar Spring, was a stage stop on the Wells Fargo line running to Westgate from Belmont. The stop was established in May 1867 and was later renamed Baxter after a local resident. Little was built at Baxter, and the *Belmont Courier* derisively referred to the "huge"

city of Baxter in the 1870s and 1880s. The small settlement never amounted to much until Davenport, Ratschki, Allen, and Church discovered gold in late 1905 in the hills just west of the old stage station.

A small tent camp sprang up almost immediately as the rush to the area began. In February 1906 George Wingfield bought the Baxter Springs mine for $25,000. By the end of the month 200 people were at the camp, and the Baxter Springs–Manhattan Mining Company, of which Thomas Edwards was president, began working some new mines, including the St. Paul, the Golden West, the Mountain Maid, and the OK. By the spring of 1906 the population of the camp had peaked at 400. Business establishments in the town were mostly in tents and included four saloons, two grocery stores, two general mercantiles, a couple of boardinghouses, a hardware store, and a few restaurants. By March a tri-weekly stage ran to Tonopah, and a dozen frame structures had been completed. Unfortunately, the ore veins were extremely shallow and quickly ran out. By the end of 1906, in less than a year, the town was abandoned. A last gasp was heard when H. L. Davenport, who had put up the second tent in town, made a strike in his Dutch mine, but the vein pinched out, and he left. The only other mining activity began in 1940 when Frank Cocca discovered a deposit of gold and silver ore that assayed at $53 a ton. The seventy-four-year-old Cocca began to develop the mine but died while working in it in June 1941. No other activity has taken place in the area since.

Although there aren't any buildings left standing at Baxter Spring, a number of scattered stone ruins and foundations remain near the few mines of the district. Nothing at all is left of the old stage station. Caution: the road to Baxter Spring is extremely rough.

Belmont

DIRECTIONS: From Tonopah, head east on US 6 for 5½ miles. Take Nevada 376 north for 13.3 miles. Exit right onto old Nevada 82 for 28 miles to Belmont.

Belmont is the queen of Nye County's ghost towns. Its history began in October 1865 when an Indian discovered a rich deposit of silver in the Toquima Mountains. Soon afterwards the newly formed Combination Mining Company purchased this discovery—the Highbridge mine. A small camp called Belmont (meaning "beautiful mountains") soon formed. By the beginning of 1866 a full-scale rush had begun. Between 1866 and 1867 Belmont was credited with having a population of up to 10,000, but 4,000 is a better estimate of the number of people living in the bustling town and an adjacent but separate camp, East Belmont, which also contained a Chinatown. The Belmont boom drained the population from many nearby towns, including the Nye County seat, Ione. As Belmont continued to grow, the residents began to call for a county seat change. In February 1867 Belmont was selected as the new county seat. The town was appropriated $3,400

The Cosmopolitan Dance Hall and Saloon, once one of the most photographed buildings in Nevada, was destroyed by vandals in 1989. (Shawn Hall collection, Nevada State Museum)

The Belmont Courthouse is currently undergoing a painstaking restoration. (Shawn Hall collection, Nevada State Museum)

to construct a courthouse. Construction began on a huge, two-story brick building but wasn't completed until 1874. A post office, with Lucius Moore as postmaster, opened in Belmont on April 10, 1867, and the town seemed destined for permanence. By 1868 businesses in Belmont and East Belmont included the Highbridge Hotel, owned by B. McCann; the U.S. Coffee and Oyster Saloon, owned by C. R. Lamont; the San Francisco Restaurant, owned by Michael Leach; the Cosmopolitan Saloon, owned by Mart Carlbol and George Fridham; the National Bank; and the Combination Hotel, owned by McCornell and Quilles. In addition, there were close to 100 other stores and saloons in the town. McClutheon and Addington ran the Pahranagat Stage, which went from Belmont to Hiko. The Belmont Hook and Ladder Company formed, but it had little to do with fire fighting and was really a social club.

In Belmont's early years mining in the area was very successful. Miners were working ten major mines within a year after the first discoveries had been made. The deepest of these was the 500-foot-deep Belmont mine, which also had 2,000 feet of lateral workings. The full potential of the mine could never be realized because of the exceptionally heavy flow of water that continually hampered mining operations. Other important mines were the

Monitor-Belmont, the Arizona, the Combination, the Highbridge, and the Green and Oder.

Six mills operated in and around Belmont during its peak years. The first, built in 1866, was a ten-stamp mill, which continued to operate until 1869. In 1867 a larger, twenty-stamp mill, the Highbridge, began operating. This mill operated for only a short while before being moved to Gold Mountain in Esmeralda County in 1880. Belmont's largest mill, the Combination, started up in February 1868. It had forty stamps and cost more than $225,000 to erect. There were also three quartz stamp mills—a five-stamp, a twenty-stamp, and a thirty-stamp. The twenty-stamp mill was the Monitor-Belmont, which the Belmont Silver Mining Company built in 1867. The thirty-stamp El Dorado, also known as Coover's mill, treated ore from the El Dorado South mine. Colonel David Buel, who was prominent in the development of mines at Austin and Eureka, was the head of the Belmont Mining Company, which owned the Buel, the Gilleland, the Transylvania, and the El Dorado mines. In April 1867 he sold the properties to J.W. Gashwiler, M.J. McDonald, and S.M. Buck, who formed the company and retained Buel as superintendent and president. There were also five sawmills in the area, all of which were extremely busy turning out board lumber for buildings at Belmont. The largest

Rear view of the courthouse, showing the locations of the jail cells pulled out for use in the Gabbs jail. The jail cells were recently returned and sit adjacent, waiting to be reinstalled during the restoration. (Shawn Hall collection, Nevada State Museum)

of these, owned by men named Crowell and Myers, produced more than 4,000 board feet of lumber daily.

Newspapers were an integral part of Belmont. There were three of them during the town's early years: the *Silver Bend Reporter,* the *Mountain Champion,* and the *Belmont Courier.* The *Silver Bend Reporter* was the first Belmont newspaper, and its debut issue was published on March 30, 1867. Oscar Fairchild ran the paper, and his brother, Mahlon, was editor. The paper started out as a weekly but eventually became a semiweekly published on Wednesdays and Saturdays. Things did not work out for the paper, however. It folded on July 29, 1868, and was moved to Austin. The editor was disgusted with Belmont and let it be known in the final issue: "The local support received by the *Reporter* has amounted to but a moiety of what it should have been and with business as it now is at Belmont, even the local columns of a weekly newspaper can contain but meager scraps of information swelled to importance only by an imaginative brain and painted by some valuable pen." He added, angrily, that Austin had always been the true center of attention and that he would move the paper there. His attitude is strange, since the paper was crammed full of advertisements for local businesses.

The second paper in Belmont was organized solely as a political weapon. Thomas Fitch, known far and wide as the "silver-tongued orator of the West," owned the *Mountain Champion.* Fitch was running for a congressional seat and felt that having his own newspaper would give him an advantage in the election. The first issue of the weekly tabloid came out on June 3, 1868. The paper only lasted until shortly after Fitch was successfully elected to office. The editor of the paper, Edward McElwain, suspended the paper on April 24, 1869, and he moved to Shermantown (in White Pine County), where he started the *White Pine Telegram.*

The *Belmont Courier* was first published on February 14, 1874. John Booth was the owner. Booth, along with Andrew Casamayou, put together one of the best-liked and most successful newspapers ever published in Nye County. Things went smoothly for the paper until 1875, when Casamayou died. After Casamayou's death on November 30, Booth lost most of his enthusiasm for the paper and finally decided to leave it on December 16, 1876. Andrew Maute took over, and Sam Donald joined him in 1880. By then Belmont was well on its way to becoming a ghost town, and the paper had a rough time. Donald left in 1889, leaving Maute to keep the struggling paper afloat. Maute leased the paper to F. G. Humphrey in 1898 after Maute was elected Nevada State Printer. The *Courier* finally folded on March 2, 1901. It published the following closing statement: "Our last issue—every branch of business in Nye County is dull."

Belmont was not always dull. There were a number of murders and even a couple of lynchings in the town. The first major conflict occurred as a result of dissension between Irish and Cornish miners. It was April 17, 1867, when

the Irish miners marched on the Silver Bend Mining Company's offices. They took the president—R. B. Canfield—placed him on a rail, and paraded him around town, stopping at most of the saloons. The group became increasingly mean as they continued to drink. Louis Bodrow, a former Austin marshall, dared to confront the mob. John Dignon, one of the marchers, hit Bodrow, and a gun battle ensued in which Bodrow and Dignon were killed, a number of men were injured, and Canfield escaped.

The second incident involved the lynching of two men, and it remains a black mark in Belmont's history. The two, Charlie McIntyre and Jack Walker, had been involved in a shooting in May 1874, and both had been arrested. They escaped but were soon found hiding in an old mine shaft. That night vigilantes lynched the pair. The man that McIntyre and Walker had killed had been a troublemaker in Belmont, and the circumstances of the death are unclear. Memories of the night still linger at the town.

After a slowdown from 1868 to 1873 Belmont received a big boost when a number of rich new deposits were discovered in the Belmont, the Highbridge, the Monitor-Belmont, and the Quintero mines in 1874. Most of these mines were located about a mile east of Belmont, near the Combination mill, where a small settlement known as East Belmont had formed a few years earlier. Belmont soon had a population of 2,000 again, and the outlook seemed good. Businesses in Belmont and East Belmont in 1874 included Vollmer Brothers General Merchandise; the Belmont Drugstore, owned by R. M. King; the Silver Bend Livery Stable, owned by W. L. Plumb; the News Depot,

owned by Granger and Black; the East Belmont Market, owned by E.C. Lead-beater; the Ephrain and Scalig Store; the D.A. Hopkins Store; the Esser and Stimler Variety Store; the Cosmopolitan Saloon, owned by J.R. Seymour; the Post Office Saloon, owned by Carpenter and Seymour; the Belmont Brewery; S. Tallman General Merchandise; the East Belmont Saloon, owned by Cravens and Mitchell; the Canfield Boardinghouse, owned by W.V. Price; the Belmont Lodging House, owned by Thomas Warburton; the Huey and Mead Lumber Yard; the Pioneer Market, owned by C.H. Hatch; the Washington Brewery, owned by George Curschman; the Franco-American Restaurant, owned by Rigant, Mettetal, and Cartier; the Lafayette Restaurant, owned by Dugnat and Jacquier; and the Star Restaurant, owned by Louis Fidanza. Fraternal organizations were also prominent in Belmont. They included the International Order of Oddfellows (I.O.O.F.), which had a meeting hall, and the Knights of the Ancient Universal Brotherhood. S. Grant Moore and W.N. Tourdrow were the local doctors, and J.A. Ball was the undertaker. Religion came to Belmont with the completion of St. Stephens Episcopal Church.

Cluggage set up a stage line to Austin, and in September 1876 a telegraph line to Eureka was completed. Western Union hosted a big celebration that culminated in a 100-gun salute. However, the excitement was tempered by the fact that all the mills in town were idle. In July 1878 the newly formed Highbridge Consolidated Silver Mining Company took over the Combination mine and mill from Abel Bennett and Joseph Brown. The company renamed the mill the Highbridge. That this transfer and renaming occurred is seldom noted, leading to confusion over which mill was the Highbridge. The old Highbridge mill, dating from 1867, was a separate mill with a different owner. The mill ruins currently labelled on maps as the Highbridge are actually ruins of the Cameron, which was built with bricks from the dismantled Combination, then Highbridge, mill in 1915.

In September 1878 the El Dorado South Mining Company, a strong producer in the 1870s, suffered a fatal blow when the mill and hoisting works were set on fire and destroyed. Both were uninsured, and the $66,900 loss forced the company into bankruptcy. The town received a boost in December, when the Highbridge Company completed repairs on the old Combination mill and restarted twenty-five stamps. Unfortunately, however, by October 1879 the mill was idle again. In January 1880 the Belmont Mining Company started a ten-stamp concentrating mill, and by summer the company was the only active entity in the district.

Beginning in 1880 the big boom at Gold Mountain, near Bonnie Clare, drained many resources from Belmont. Equipment from the old Highbridge and Combination mills was moved to Gold Mountain, as were many vacant buildings. Despite these blows, the Monitor-Belmont mill restarted in June 1881, but the lack of consistent ore shipments led to the closure of the Wells Fargo office in 1884. By 1885 only limited activity was taking place in Belmont. Miners worked only the Highbridge mine, owned by John Griffin and Casper Piel, and the Arizona mine, owned by John Delanber. The sale of the Belmont Mining Company's property in 1887 was the death knell for Belmont's mining. The Monitor-Belmont mill was still running but on ore from Barcelona, not from Belmont. By the end of the year even the mill had closed. During the 1874 to 1887 revival Belmont's mines had produced an additional $2 million worth of ore. From 1865 to 1887 their recorded production was valued at over $15 million.

By 1889 many of Belmont's businesses and most of its residents had left. The few businesses remaining included the Ball and Deady Drugstore, the Belmont Brewery and Saloon, Warburton's General Merchandise, and Ernst and Esser General Merchandise. The Tate and Wallace Belmont-to-Sodaville stage and Tate's Austin-Belmont stage were still running, but they folded the following year. For the most part, the mines remained quiet until about ten years after the turn of the century. Following a six-year hiatus, Thomas Warburton and W. A. Atwell restarted the Monitor-Belmont mill in 1893 and worked old tailings from dumps in East Belmont, but this activity only lasted a couple of years. Most of the old mills of Belmont were torn apart and moved to other locations in the 1890s. John Mayette took the old ten-stamp battery of the Belmont Mining Company to Union, and sections of other mills were used to construct a mill at New Pass in Lander County.

The final concession of defeat came from the *Belmont Courier* in its last issue on March 2, 1901:

With this issue, the *Belmont Courier* suspends publication. For 28 years, we championed every cause that tended to help Nye County in particular and the State of Nevada in general and it is with keen regret that its proprietor calls it from the field of battle. But every branch of business in Nye County is dull and for several years, the *Courier* has brought in very little money and at the present time it is not a paying institution. From present appearances, it will be a hard matter to make a newspaper pay in Nye County for a long time and as the lessee of the *Courier* has a chance to enter into a more remunerative business enterprise, he has decided to do so. The patrons of the *Courier* have our sincere thanks for what they have done toward making it possible for the paper to live for so long a period. We wish our readers goodbye and hope that one and all will be happy and prosperous.

The paper's owner left and began running a stage line from Tonopah to Sodaville. This was the end of the most important early newspaper in Nye County. By the time it was about to cease publication, not one business in Belmont was advertising in the paper.

When Jim Butler left his Monitor Valley ranch in May 1900 to head for the strikes in Klondike (in Esmeralda County), no one knew his trip would change the state's history. After Butler discovered rich silver deposits at Tonopah Springs, a huge rush began to that area, emptying many small towns near the site. Almost immediately a strong call went up to transfer the county seat from Belmont to the new town of Tonopah. By 1903 Belmont only had thirty-six qualified voters and could manage but feeble resistance, and in May 1905 the transfer was made. By 1911 Belmont's population had shrunk to less than fifty, and on May 31, 1911, the town lost its post office. In October 1908 the Belmont Security and Development Company bought the old Combination mill and by the end of the year had also purchased all the old mines. Unfortunately, the company went bankrupt before starting any production.

Belmont had a fairly active revival beginning in 1914. The Monitor-Belmont Mining Company, of which George Nelson was president, had acquired almost all the old mines near Belmont and in August 1915 started a huge mill, built in East Belmont, to process the ore. The mill, the Cameron, had ten 1,600-pound stamps and a 150-ton oil flotation system to process ore from the Monitor-Belmont's twenty-one claims. The bricks used to construct this mill were taken from the Combination mill, which had been torn down the year before. The Cameron mill, named for the company's superintendent, was also used to rework some tailings from earlier activity. A $15,000 power line was built to Manhattan to provide power for the mill and electricity for the town. The company, which employed thirty, had three major mines, which kept the mill running until 1917. In February the Highbridge mine was connected to the mill by an eighteen-inch gauge railroad, and the company opened a clubhouse and recreation hall in Belmont. Thanks to all this activity, the population of Belmont began to increase slightly. The post office reopened on September 27, 1915, and it looked as if Belmont was going to get a second chance.

The Monitor-Belmont Company left the district after closing the Cameron mill in 1917. In early 1918 the property was leased to the Nevada Wonder Mining Company, which reopened the Monitor-Belmont mine. Nevada Wonder spent huge amounts of money on exploration that turned out to be fruitless, and the company relinquished its lease in 1919. But Belmont was not yet ready to die. A thirty-ton cyanide plant opened in 1921 to treat the old tailings. When it shut down Belmont drifted back into ghosthood. The post office closed on August 31, 1922, and there have been no revivals since. The activity that took place from 1914 to 1922 yielded a reported $1 million worth of ore. The town still had twenty-eight residents in 1945, but soon afterwards that figure was down to ten. In 1954 and 1955 there was a

Belmont in 1893 after the town had begun to decline. The Monitor-Belmont Mill is visible in the far left background. (William Metscher collection, Central Nevada Historical Society)

flurry of interest over uranium in the area. The Gold Metals Consolidated Mines Company, the Red Hill Florence Mining Company, the West Uranium Mines Corporation, and others filed 108 claims, but while some uranium was produced, the quality was poor, and activity soon ended.

In the author's opinion, Belmont is one of the very top ghost towns in the state. The remains at and around the site are amazing. There are the picturesque ruins of the three mills. Only a stack and some rubble mark the site of the Monitor-Belmont mill, located just below the town. The Combination mill is marked by a huge, pockmarked stack and extensive ruins, including a small brick room that housed explosives. The best of the mills is the Cameron, whose enormous, skeletal brick walls are awesome to behold. Both the Combination and the Cameron mills are in East Belmont, a mile east of Belmont. Recent maps and a Forest Service sign near the Cameron mill have mistakenly labeled the mill as the Highbridge, an error that has been repeated in a few recent books about Nevada. The reason for this error is discussed earlier in this section.

The remains at East Belmont are scattered over a wide area. Behind the Cameron mill ruins are about thirty stone cabins in various stages of decay. This section appears to be part of the old Chinatown. Many other stone ruins are scattered along the main road running from East Belmont to Belmont. The old horse racetrack is still faintly visible near the Combination mill ruins. The track was known as Monitor Park and operated for a number of years in the 1860s and 1870s.

Belmont itself remains an incredible site. Although a number of newer homes have been built in the town, mainly by residents of Tonopah and Las Vegas, the grand beauty of "old" Belmont shines through. First, there is the beautiful and imposing courthouse, which is undergoing a slow, painstaking restoration. The Nevada State Parks System, which now controls the building, has protected it against deterioration and vandalism by putting plastic over the windows and installing locks on the doors. Portions of the first floor have been partially restored, and small displays have been put together. During the summer months the courthouse is opened for public viewing. After the building was abandoned it was used for many years as a place to store hay for one of the nearby ranches. The cells of the jail, built in the back section of the courthouse, were torn out and used in the Gabbs jail. Thanks to the efforts of the Metscher brothers of Tonopah, the cells were returned to Belmont recently when Gabbs constructed a new jail. The old cells remain behind the courthouse awaiting eventual placement in their original home.

The main street of Belmont contains some of the best remains the author has ever seen. In addition to the ruins of a few dozen buildings, there are quite a number of structures still standing, some of which have recently been restored. One of the most impressive of these is the old stone offices of the Combination Mining Company, located at the north edge of town.

The building, now used as a residence, was restored in the 1970s. However, a tragic event has deprived visitors of one of the most remembered and photographed buildings in any Nevada ghost town. The Cosmopolitan Dance Hall and Saloon had been a landmark in Belmont for 100 years. It was at the Cosmopolitan that the famous actress Lotta Crabtree performed in *Uncle Tom's Cabin*. Unfortunately, a group of raucous four-wheel-drive enthusiasts decided to see if they could pull down the building. After attaching chains and ropes to its main supports, they only had to pull a couple of times, and the building came crashing down. A senseless and stupid act has deprived historians and tourists of a valuable part of central Nevada's heritage. What makes the loss even more devastating is that plans had been made for a partial restoration of the building. Now, only a pile of wood rubble is left of the Cosmopolitan.

Besides its multitude of ruins, buildings, and mills, Belmont has one of the most extensive and fascinating cemeteries in the state. The large cemetery, located just south of the Monitor-Belmont mill ruins on the east side of the road, includes many wooden head boards, fancy wrought-iron fences, and elaborate gravestones. Jack Longstreet and his Indian bride are both buried in the cemetery.

For the past seventeen years, the author has found himself constantly going back to Belmont for just one more look. Plan on spending a day or two to completely tour Belmont and East Belmont. Before traveling to Belmont, stock up on necessary supplies such as food and gas, because they aren't available at the site. However, Dick's Belmont Bar is normally open on weekends to quench any thirst. Don't miss Belmont!

Box Springs

DIRECTIONS: *Box Springs is 8 miles south of Wilson's.*

Box Springs was a horse-changing stop on the Eureka–Belmont wagon road beginning in 1875. When the Belmont revival faded in the mid-1880s, the line and station were abandoned. Nothing remains at the site.

Breyfogle

DIRECTIONS: *Breyfogle is located 3 miles east of East Manhattan.*

In May 1906 John Zabriskie made some gold discoveries, and a small camp, named Breyfogle because he believed that the discovery was the lost Breyfogle mine, immediately formed. When nobody found much else in the area the camp disappeared just as quickly. Over the next twenty years some exploration and prospecting took place, but the mine produced nothing. Prospectors built a couple of cabins at Breyfogle, and they remain in the canyon today.

Central

DIRECTIONS: *Central is located 2½ miles west of Manhattan on the south side of Nevada 377.*

Central was one of many small camps that sprang up in the Manhattan district in the early 1900s. It was the largest and best established of these minor camps, even though it lasted less than a year. Prospectors found gold there in the spring of 1906, and a small tent camp quickly formed. Only a month after miners had made the initial discoveries, more permanent wood buildings—including five saloons, two stores, two hotels, an assay office, a lumberyard, and a bakery—were being built. Residents planned to construct an electric light plant, but after prospectors made new strikes in Manhattan that town became the focus of attention, and the plans were dropped. A post office, with L. H. Glar as postmaster, opened at Central on March 22, 1906, but it closed on September 19 as the town began its quick collapse. At its peak, Central's population was about 100. Everyone living in Central moved to Manhattan in late summer, soon after the latter town's revival began. Most of Central's buildings also moved to Manhattan, and today only small mine dumps and some scattered wood mark the site of the once-bustling town.

Darrough Hot Springs

(Hot Springs)

DIRECTIONS: *Darrough Hot Springs is located ¼ mile off Nevada 376, 8 miles north of the junction with Nevada 378 at Round Mountain.*

Indians used the hot springs located at Darrough Hot Springs for many centuries before white people came to Nevada. John C. Frémont visited the site in 1845 during his fabled exploits. A stage station, the first in the Big Smoky Valley, was built in the 1860s at the springs to serve the Belmont–Austin line beginning in the 1860s. James T. Darrough bought the stage station and hot springs in the early 1880s and gave the area his name. Later he built a hotel, which operated until the teens, thanks mainly to the attraction of the hot springs. The family house, however, burned in 1909. Darrough remained at Darrough Hot Springs until he died in 1911 at the age of fifty-two. His wife, Laura, lived to the age of ninety-two. She died in 1952 and was buried next to her husband at the family cemetery near the springs.

The station and immediate grounds remain in the possession of Darrough's descendants. The old stone stage station still stands and is in excellent condition. A small wooden addition has been added to serve as a boardinghouse. The hot springs still run. A pool has been installed, and visitors can swim in the mineralized water for a small fee.

East Manhattan

DIRECTIONS: *From Manhattan, take Nevada 377 east for 2 miles. Exit right onto a poor road and follow it for ¼ mile to East Manhattan. Signs mark the entrance to the road.*

The small settlement of East Manhattan sprang up in early 1906, soon after Manhattan began booming. Manhattan's richness prompted a great deal of exploration in the area around the town. Prospectors located a few promising ledges about two miles east of Manhattan in January. Soon a camp of seventy-five had formed around springs near the budding mines, and the Nevada Brokerage Company laid out the East Manhattan townsite. East Manhattan's main mine was the Buffalo. Its owners, Elwood Maden and Omer Maris, sold out to the Buffalo Mining Company. Other mines opened were the Mammoth, the Consolidated, the Mineral Hill, the Manhattan St. Paul, the Manhattan Minneapolis, the Gold Ridge, the Pine Nut, and the Crater. At its height during the spring of 1906 East Manhattan had two stores, two saloons, and a restaurant, all housed in large tents and one wood building. The Toquima Copper Company and the Bonanza Copper Company both began operations in the area.

Activity had ceased completely by the end of 1906 because the ore veins were very shallow and quickly ran out. The tent camp folded, and soon everyone was gone. The final attempt to keep East Manhattan going came in 1911 when the Mineral Hill Mining Company, the St. Paul Mining Company, and the Minneapolis Mining Company consolidated to form the Mineral Hill Consolidated Mining Company. But the new company was unable to locate any substantial new deposits, and it soon folded. Not many signs of the settlement remain, but a nice spring marks the site of the camp. Less than a quarter of a mile east of the spring are the mines, which haven't been worked since 1906. The mines were re-timbered in the 1950s, but no production was recorded. The shaft and hoist house are still in good shape, and the original powder house is located nearby.

Georges Canyon

(Fresno) (Marsh)

DIRECTIONS: *From Stone Cabin, head west for 1 mile. Bear left and continue for 2 miles to Georges Canyon.*

Limited mining activity took place in Georges Canyon from 1903 to 1931. In 1903 Ira Farcher discovered gold and silver ore in the canyon that assayed as high as $300 per ton. Work soon started on a mine named the Clipper. The Fresno Mining District formed in June 1909 after Farcher and Harry Stimler found a new ore deposit. A small tent camp of ten, named Fresno, formed near the mine and mill. Farcher's brother, Ira, discovered placer gold in July. In July Stimler also hit a rich vein with initial assays of $1,500 a ton and organized the Georges Canyon Mines Company, which also owned the Tiger and the Bernice mines at Ellendale. As a result of this activity, William Mikulich platted a townsite and named it Marsh after William Marsh, one of the town's first residents who, along with Harry Stimler, later became one of the early discoverers of ore at Goldfield. By the end of August sixty people were living in the district. In October Jack and Frank Meyers struck ore that assayed at $300 a ton, and the first shipment of ore from the camp was sent to Tonopah. Activity continued unabated in 1910. The Farchers struck another high-grade vein, Grant Murphy worked the Golden Anchor, and the Marsh Mining and Milling Company, of which J. H. Warburton was president, formed in November. In April 1911 Stimler and Marsh built a two-stamp mill on the Red Ruth property, but it was used primarily to prepare local ore for shipment. However, interest in Georges Canyon faded during the following year, and by 1912 only the Farcher brothers, Stimler, and Marsh were left in the district. The Farchers continued to work their mines for many years and in 1929 built a two-stamp mill nearby. They stopped work on the mine in 1931 after it had produced a little more than $12,000 worth of gold and silver. When operations were curtailed the shaft was 180 feet deep.

Joe Clifford, who at that time ran the Stone Cabin Ranch, also did some mining work in Georges Canyon. Clifford owned the Little Joe claim, whose ore assayed at $20 per ton. When work stopped on the claim after a few years, the shaft was almost 90 feet deep. Georges Canyon experienced a small-scale revival in late 1928 when employees of the Tonopah Extension Mining Company began reopening some of the mines. Charles and William Farcher returned to work their mine, and William Woods and Jack Jordan were shipping $35-a-ton ore to Millers in Esmeralda County. But this burst of activity had ended by 1931. Georges Canyon last saw any mining activity when the Ben Hur Mining Company started operations in 1933. The company installed a small, six-ton pilot mill in 1934, but, after failing to find satisfactory ore, the company ceased to operate. There are few remains left to show that activity ever took place in the canyon. One stone building and mill ruins are the only markers.

Gold Hill

DIRECTIONS: From Round Mountain, take a dirt road north for 4 miles to Gold Hill.

Gold Hill was never a settlement or town but was rather a small mining complex that formed after Charles Peraine discovered gold in June 1924. Intensive mining operations began there in 1927 when the Gold Hill Mining Company purchased the Nonpareil group of claims, which included the Gold Hill mine. The company, incorporated in November 1927 with $1 million in backing capital, built a small stamp mill with cyanide tanks the following year. The Gold Hill Company was reorganized in April 1929 and was renamed the Gold Hill Development Company. In 1930 the company was once again reorganized and renamed, becoming the Gold Hill Consolidated Mines Company. P.H. Murray of Visalia, California, was president, and W.H. Farris of Round Mountain was mine superintendent. The company was controlled by the Tonopah Mining Company and the Tonopah-Belmont Mining Company. Equipment at the mine included a fifteen-horsepower Fairbanks-Morse hoist and a two-drill compressor.

Another company, the Gold Zone Divide Mining Company, was also active near Gold Hill during the early 1930s. The company was originally called the Tonopah Gold Zone Mining Company, but it underwent a name change when it was reorganized and incorporated in February 1918. In late 1929 the company purchased the mines controlled by the Divide Mining Company near Gold Hill and sank a shaft in 1930. The company found decent gold ore, but it was shallow. The Gold Zone Company left the Round Mountain district in 1933 to concentrate on its holdings in the Divide district, south of Tonopah.

In August 1930 a $770,000, 100-ton mill began operating on property near Gold Hill that the Tonopah Mining Company and the Tonopah-Belmont Development Company had purchased in 1929. The Gold Hill Development Company formed to run the operation, and it quickly increased the mill's capacity to 125 tons. By 1931 the company was producing $10,000 worth or ore a month. But by 1933 ore values began to fall, and all operations were curtailed in March. In 1934 T.F. Cole bought the mill and the surrounding property for $25,000, but he died in 1938 before any real plans could be carried out. Miners worked the Gold Hill area periodically after his death, but not to any great extent. Production value from 1930 to 1942 for the Gold

Hill mine and other nearby operations stands at just under $1 million. The mine hoist remains at the site, along with the foundations of the mill.

Hannapah
(Silverzone) (Silver Glance) (Volcano) (Bannock) (Sylvanite)

DIRECTIONS: *From Tonopah, head east on US 6 for 18 miles. Exit left and follow the road for 1½ miles to Hannapah.*

The Graham brothers made initial discoveries in the Hannapah area in 1902. Soon the Hannapah Mining Company, with Frank Work as president, began work on the Hannapah mine, and by 1905 a small camp had formed nearby. Prospectors explored the district further during the next few years. In January 1906 miners made a new strike a short distance away from the Hannapah mine. It attracted considerable attention. A townsite called Silver Glance was platted in early 1906, and on February 6 the promoters—Thomas Griffin, Dave Holland, and Harry Chickering—hosted a huge barbecue to attract potential customers to the new district. The people liked the barbecue but not the townsite, and they bought very few lots. The Silver Extension Mining Company, of which R. B. Davis was president, worked the mine, but by 1907, after only limited mining production had taken place, the district and townsite were abandoned. In July the mine, the hoist, and the buildings of the Hannapah Mining Company were destroyed by fire.

In 1913 George Rodgers and Charles Glenn discovered silver just northwest of the old Silver Glance townsite. A short boom followed. The small settlement that sprang up at Cedar Corral was called Volcano. Soon afterwards Johnny Musser and Charles Glenn's new discoveries created a rival camp, called Sylvanite, about a mile to the northwest. Excitement reached a fever pitch in April 1915 when Nick Abelman found ore that assayed at $11,000 a ton. By May 100 sacks of ore had been shipped from Sylvanite. In July Glenn made the first ore shipment from Volcano. It returned $213 a ton. But by 1917 the ore pockets at both camps had run out, and both were abandoned.

The Hannapah mine reopened in 1915, and Frederick Browne, who owned it, claimed that $150,000 worth of ore had been exposed when the mine was retimbered. The mine was known to be rich, but earlier operations had removed the easily mined richer ore, and it was believed that further exploration of the property would be unprofitable. Browne proved this assumption incorrect, and the Hannapah mine produced from 1919 until early 1921.

Two other mining companies were active in the district during the 1919

revival. One was the Silverzone Mohawk Company. William Gray and J.J. Clark, both from Tonopah, were the owners. They also owned the Silverzone Mines Company, which worked property in Silver Glance, a quarter of a mile west of Hannapah. The Silverzone Mohawk Company sank a new eight-foot shaft that yielded ore assaying at an average of $30 per ton. The company leased the old Tripod mine, which miners had originally worked a little bit in 1905, to Charles Spilman, and ore removed from the mine assayed as high as $80 per ton.

The second active company in Hannapah was the Hannapah Divide Extension Mines Company, which was incorporated in December 1919. W.R. Porter was president. The company owned nine claims in the Hannapah district, now known as the Silverzone district. The company reopened old shafts in 1919 and sent small shipments of silver ore to mills in Tonopah in 1920. In April 1919 Senator Roy Thatcher from Eureka bought the Gold Horn mine for $20,000, and Charles Topping bought the Silver Leaf for $35,000. Jack Clark, who had been at Hannapah since 1902, sold both of the mines. In April 1920 E.P. Cullinan bought all the claims at Sylvanite and Volcano for $25,000.

By 1921 the district was dead once again except for some claims that remained active. Ben Richardson, who had bought a number of claims, including the Hannapah, the Silver Glance, the Silver Trumpet, and the Running Elk, began exploratory work in 1922. He was unsuccessful, but the Hannapah Extension Mines Company decided to gamble and purchased the claim from him in 1922. The company lost the gamble when the ore turned out to be quite poor in quality. Richardson continued to work his other

claims, finding a vein rich enough to warrant digging a shaft well over 300 feet. He built a small Straub mill with an ore house to process ore from his mine. He continued to work the mine until 1935, when the ore ran out and he left the district.

The last revival in the Hannapah district took place in 1927, when three companies became active. In that year, the World Exploration Company of Fort Worth, Texas, purchased the Hannapah Extension Mines Company, which had been idle since 1922, and began limited activity that lasted until 1929. It purchased the claims from Fred Browne, a longtime Hannapah resident, and he was named general manager. The company constructed a number of buildings to house the miners, but financial problems then forced it into bankruptcy. The second company in the district was the Apache Hannapah Mines Company, which was incorporated in 1927. Walter Lynch of Alhambra, California, was president. The company owned four claims in the district; they were the Helen, the Helen Number 1, the Silver Horde, and the Dickey Bill. These claims had values in both gold and silver. In 1928 the company sank a new 300-foot shaft. It strung an electric line from Tonopah to power the electric mine hoists, but the venture proved fruitless, and the company folded in 1929.

The last company to work the district was the Hannapah Silver Star Mining Company, which organized in early 1928 with Russell Boardmen as president. The company owned seven claims in the Hannapah district, all of which were adjacent to the Hannapah Extension mine. But the company struck water in all the deep mines, which stopped most mining. Some activity continued, without much profit, until early 1929. When the company folded, the Hannapah district joined the long list of Nye County ghosts. Only the

Richardson family and Jack Clark, who had lived at Hannapah for twenty-five years, stayed on, hoping to find another rich deposit. In 1934 Clark started a ten-ton mill to rework some of the old tailings and new ore from his claims. The following year Richardson also started a small Straub mill, but in September he died of pneumonia after he was seriously injured in the mill. By 1940 Clark had left the district, which was empty until 1955 when the Uranium and Federated Minerals Company reopened Richardson's old Silver Leaf mine. But it gave up in 1964.

Very little is left of Hannapah. A careful search reveals the original site and remains of one wood building just below the Hannapah mine. There are also a few other wood ruins along with the complete workings of two mines, including the mine hoists and engine housings. Some ruins mark the Silver Glance townsite, while only small mine dumps are left at Sylvanite and Volcano. Exercise extreme caution when visiting Hannapah, for unmarked shafts are located throughout the site and are well hidden.

Indian Springs

DIRECTIONS: From Manhattan, head west on Nevada 377 for 6.8 miles. At the junction with Nevada 376, exit left (south) and follow it for 2.8 miles. Exit right and follow the road for 2 miles to Indian Springs.

The small camp of Indian Springs formed in 1865 around a mill just to the north of San Antonio. A ten-stamp mill was built in 1865 to treat the extremely high-value ore coming out of the nearby San Antonio Mountains. At one time, the camp had a population of ten. Indian Springs ceased to exist when ore values declined and the Pioneer mill was shut down and moved to Northumberland in 1868. Today only scant rubble marks the site.

Jefferson
(Millville)

DIRECTIONS: From Round Mountain, head north out of town for 1 mile. At the fork, take a right and follow the road for 1½ miles to a point overlooking a small canyon with a stream. Continue straight (east), parallel to the stream, for 4 miles to Jefferson. There are a few stream crossings, but none are very deep. Ruins of Jefferson are scattered along the canyon for ¾ mile.

R. Chanrock and A.V. Wilson made the initial silver ore discovery—the Silver Point—in Jefferson Canyon in August 1865. Prospectors made more substantial discoveries in 1871, when a test load of silver ore sent to Austin returned over $28,000 worth of silver. In 1873 two men, John Johnson and Robert Fergerson, discovered rich silver ore, and the Green Isle Mining District formed. Two principal mines, the Prussian and the South Prussian (also known as the Jefferson), started operations. Charles Kanrohat discovered another mine, the Sierra Nevada, which also became a fairly consistent producer. A large ten-stamp mill was built at the Jefferson mine in 1874, and another ten-stamp mill was built at the Prussian mine soon after. The Prussian mine, which had a 250-foot shaft, constantly had water trouble that hampered operations. By 1874 other active mines in the area were the Sailor Boy, the Prussian Troy, the Keokuk, the Buckeye, and the Hillside.

It was not until construction of the two stamp mills had been completed that interest in the Jefferson area increased. By the end of the summer of 1874 the town contained two separate sections. The main city district was three quarters of a mile above the mining and milling area. It included a post office, with George Kilbourne as postmaster, which opened on October 22; a Wells Fargo office; the Dayton stone store; the Ferguson Brothers Hotel, which was later moved to Lower Town; seven saloons; three restaurants; two stables; two bakeries; a brewery; a butcher shop; a barbershop; a lumberyard; and 100 homes, 20 of which were frame homes. Its population was 250. Lower Town included both mills; the Price and Lamb Hotel, which was also known as the Hotel de Jefferson; a boardinghouse; mine offices; and a few homes. In 1874 a toll road was constructed over the rugged Jefferson Mountain to Belmont via Meadow Canyon. Supplies for the booming town arrived daily by mule team from Wadsworth, near Reno.

The Elsa Mining Company Mill during the 1930s. (Central Nevada Historical Society)

Preserving the Glory Days

Jefferson's peak years were 1875 and 1876, and its production value during that time was almost $1.5 million. During the boom years prospectors filed 120 claims, but most turned out to be merely granite deposits. The population of Jefferson stood at a little more than 800, but by the fall of 1875 the mills were only operating sporadically, mainly on ore hauled in from Barcelona. Many families began to leave, and only five students were left in the school. In January 1879 the Jefferson Company officially folded, and by August the Belmont–Austin stage line was rerouted and no longer went through Jefferson. Soon afterwards, in 1876, production slowed, and the mill at the Jefferson mine closed. In 1878 the Prussian mine and mill followed suit. The post office closed on January 29, 1879, sounding a death knell for the town. Between 1877 and 1882 only forty tons of ore were removed from the claims, and, reportedly, only four miners remained at the town to continue operations. In January 1883 the Jefferson Silver Mining Company formed and began work on the St. Charles and Sunnyside mines. In March the company repaired and restarted the old Jefferson mill. Other mines also became active. They included the Idlewild, owned by Frank McLane; the Woodbridge, owned by Dan Deady; the Keystone, owned by McLane and James Bryson; and the Union, owned by Charles Kanrohat. Life returned to the town, and about 150 people moved back. The Wells Fargo office reopened, and a number of businesses began operating. In 1885 Charles Harrison restarted his small mill, built in 1881, but by the spring of 1885 fortunes were going bad again. On July 4 all the Jefferson Silver Mining Company's property was sold at a sheriff's sale, and the mill was quickly moved to Park

Canyon. The revival was over, and the town quickly emptied again. Another blow came in April 1886 in the form of the Prussian mill's relocation to Ophir. By 1887 virtually the only people left in Jefferson were Charles Kanrohat and the Harrison brothers. In April 1906 the Jefferson Canyon Consolidated Mining Company and the Queenie Consolidated Mining Company formed, but they had very little success before leaving.

A group of New York businessmen attempted to revive the town once again in 1908 when they purchased all the mines for $350,000. The Jefferson Mining Company formed with Charles Brenneman as president and Warren Flick as general manager. The company bought most of the mines from the old Jefferson veteran, Kanrohat. The company built a new $90,000, 100-ton mill, which it used to treat the huge mounds of ore that had been pushed aside in the hurry to get at the richer ore. At the same time, Colonel John McAllister organized the McAllister Milling and Power Syndicate. The syndicate built a mill and platted a new townsite, called Millville. These operations did not last too long, however, and soon Jefferson belonged to the ghosts again. Kanrohat was alone once more, but he continued prospecting.

Charles Stoneham, owner of the New York Giants, purchased the Sierra Nevada property in 1917. Kanrohat, who was still in Jefferson after more than forty years, sold all his holdings to Stoneham for $100,000. Production started to help satisfy the demand for metal after the onset of World War I. The mill built in 1908 was reequipped for flotation. It opened for operations despite low-grade ore, but it was unsuccessful and closed in 1917. In May Kanrohat again sold the property, this time to John Miller, John Adams, Jesse Knight, and Lester Morgan for $200,000, but they defaulted before the end of the year, and the property returned to Kanrohat. In 1918 S. H. Brady, who organized the Jefferson Gold and Silver Mining Company, purchased the old Kanrohat mine properties. The huge, 100-ton mill reopened, and the company added a cyanide settling tank. The company developed the Kanrohat mine further to a depth of more than 1,000 feet. New ore was discovered, but the size of the deposits was fairly small. The mining company's main source of revenue was the reprocessing of over 60,000 tons of ore in the old tailing dumps. During the short revival the company employed fifteen men, but it ran out of capital and folded in February 1919. Jefferson was empty once more. It saw no more activity until February 1926, when Kanrohat once again sold the property, this time to the Comstock Merger Company for $40,000.

The Elsa Mining Company made a final attempt at reviving the district when it reopened the mill and installed new equipment in 1928, but it too soon folded. The revival ended almost before it had begun, and when Charles Kanrohat died in August 1929 at the age of eighty there was no longer much interest in the district. In 1931 the Bright Star Gold Mining Corporation spent $350,000 on the development and construction of a seventy-five-ton mill, but the expense crippled the company, which folded before completing the project. The Copon Silver and Gold Mining Company began operations on the old Kanrohat property in December 1933 but was not successful. After producing $2.3 million worth of ore, the Jefferson mines closed forever.

The Jefferson ruins are extensive and extremely interesting. The lower section of the town contains ruins of the mills and quite a number of smaller buildings in various stages of decay. There are also acres of old dumps that will delight any collector. Broken glass, pieces of china, and old tin cans litter the site. A cemetery is located half a mile below Lower Town. In the upper section more substantial ruins remain; however most are remnants of the later revival attempts. Many buildings still stand in relatively good condition. The road from the lower section to the upper section is impassable, but the county recently reopened a better road that had been closed by the Forest Service.

Junction

DIRECTIONS: *From Round Mountain, head west on Nevada 378 for 2.7 miles. Exit right (north) on Nevada 376 for 27 miles to Junction.*

The Junction post office opened on March 20, 1873, at the old Lognoz Ranch on the Austin–Belmont stage line. It was the main post office for the northern part of Smoky Valley. Abraham Minnimum, who owned ranches in the area, served as the first postmaster. Minnimum later moved to a ranch bearing his name about seven miles north. Auguste Lognoz took over as postmaster in 1888. When he left in 1899 the Millett family purchased the ranch, and Christina Millett became postmaster. The post office closed July 31, 1906, after most mining activity in the surrounding area had stopped. A ranch is still active at Junction, and a few old buildings remain as relics of the early activity that took place there.

Learville
(Searville) (Learnville)

DIRECTIONS: *Learville is located ½ mile south of Northumberland.*

Little is known about this small camp that sprang up in 1868. There are a number of variations of its name, but Learville is the name that appears on postal cancellations. Learville formed around a ten-stamp mill that the Quintero Company completed in October 1868. T. F. White built the mill, and the company employed twenty men to construct it. The ten-stamp mill, which cost $50,000 to build, also had a large brick kiln. The mill employed ten men, most of whom resided in nearby Northumberland. The Quintero Company went bankrupt only two months later, and all of its property was sold at a sheriff's auction in December. A few small stone cabins were built near the mill, but none remain today. The mill itself was never a success, and after suffering through numerous problems it finally shut down for good in 1870. Some of its machinery was later used in the construction of a mill at Northumberland in 1879. The camp supported a post office from December 3, 1868, until March 19, 1869. David Rosenberg served as postmaster. A new open-pit mining operation has completely obliterated the site.

Logan Station

DIRECTIONS: *From Pine Creek, continue north for 4 miles. Exit left and follow the road for 3½ miles to Logan Station.*

In 1867 Bob Logan opened Logan Station, which served the Belmont–Austin stage line when it used an alternative route through Moores Creek rather than going through Northumberland Canyon. Once the route began to run in Smoky Valley rather than in Monitor Valley the station was no longer needed and was abandoned. Only a foundation marks the site next to Logan Spring.

Longstreet

DIRECTIONS: *From Tonopah, head east on US 6 for 18 miles. Exit left on Stone Cabin Road and follow it for 5½ miles to Stone Cabin Ranch. Take a left and follow the road for 1 mile. Take a right at the fork and follow the road for 9¼ miles to Longstreet.*

Jack Longstreet formed the small camp of Longstreet after he discovered gold and silver ore in October 1907. Longstreet's shady reputation hindered the camp's development, as skeptics worried that this was another of his numerous schemes. But in this case it turned out to be legitimate.

Longstreet, who was married to an Indian, had killed a number of men and had carved notches for each one in the handles of his guns. He wore his hair extremely long, according to him because everyone else did it but actually because in his youth his ear had been notched when he was caught rustling cattle. He lived on a ranch in Longstreet Canyon, just to the north of Windy Canyon, where the mining camp of Longstreet formed.

There were never more than fifteen people working the tunnels and vertical shafts in the canyon. The camp did not even last one year, and no

substantial buildings were erected. Joe Hutchinson and W.R. McCrea took over the property, but after a year of exploration they returned it to Longstreet and his partner, E.C. Courtney.

The district remained silent until 1928, when the Golden Lion Mining Company purchased the Longstreet mine and began fairly intensive operations. Longstreet died soon after selling the mine for $100,000. The company developed the mine to a depth of 300 feet and installed two diesel engines and a seven-drill compressor. A 100-ton cyanide mill was built in 1929, and sporadic production took place until 1931. These two years of activity produced $10,000 worth of gold and silver for the Golden Lion Company. During that time a camp of twenty-five formed, and a large false-fronted boardinghouse and half a dozen other buildings were built. The mill closed in 1931. In August Captain John Hassell bought the mine and mill and formed the Crown Reef Consolidated Gold Mines Company. The company also opened the Mockingbird mine. Since then, Longstreet has been a complete ghost town. As recently as 1940 the mill and five buildings were still standing. The mill equipment was scrapped during World War II, and the other buildings were moved away.

Longstreet's remains are not very interesting because the camp never really matured. All that is left are the mine hoists, the rubble and foundations of the mill, and one dugout structure. The road to Longstreet is extremely rough and rocky, making a four-wheel-drive vehicle a necessity. Longstreet is not worth the effort of getting there. Many old buildings and one interesting artifact — an old boiler used as a still that was split open by federal agents — remain at the Longstreet Ranch. There is potential for future mining at Longstreet. In

1985 Naneco Resources outlined an 850,000-ton body of .079-per-ton gold ore there, but as of 1995 no production had started.

Manhattan

DIRECTIONS: From Tonopah, head east on US 6 for 5½ miles. Take Nevada 376 north for 33.7 miles. Exit right on Nevada 377 for 6.8 miles to Manhattan.

The Manhattan district was active long before the town of Manhattan formed. George Nicholl discovered rich silver ore in the district in 1866. Following the discovery, prospectors located more than fifty claims in the area. Two mines, the Mohawk and the Black Hawk, were developed during the next three years. The Mohawk mine had a 100-foot tunnel, and the Black Hawk had a 60-foot shaft. The ore mined at these and other claims was sent to mills in Belmont. The ore averaged close to $100 a ton, with some of it reaching as high as $2,500 a ton. A freight line ran from Austin to Manhattan Gulch at a cost of $60 a ton. By 1869 things were slowing down, and by the beginning of winter the district was totally abandoned. In May 1877 the Manhattan Mining District was renamed the Eagle Mining District. Prospectors in the district included M. Barrows, J. Quartz, M.H. Smith, and H.L. Jones. They didn't find much, but prospecting continued. In 1885 Adam McLean, George Nicholl, John Kennedy, Alexander Carey, William McCann, and James McCafferty reorganized the district. By 1887 the Manhattan Freehold Gold and Silver Mining Company controlled the claims, but it failed to find an ore vein rich enough to warrant major development. Even though mining was sporadic, it is estimated that the district produced $200,000 worth of ore from 1866 to 1904. Manhattan remained silent until major new discoveries were made in April 1905.

A cowpuncher from the Seyler Ranch in Big Smoky Valley made the first new discovery in Manhattan. John Humphrey, Frank Humphrey, C.A. Cooper, and G.E. Maude were traveling from Belmont through Manhattan Gulch. They stopped to eat lunch, and Humphrey finished first and wandered away. He found an outcropping only 100 feet from the spot where they had been eating. The four cowpunchers managed to break off a few pieces of the ore. The subsequent assay report showed the value of the rocks to be more than $3,000 a ton. The men staked the Ida, the Lottie, the April Fool, the War Eagle, and the Tip Top claims. Following this discovery, Manhattan sprang up almost overnight. Initially a tent city of about 500, named Palo Alto, formed at the mouth of Manhattan Gulch, but residents left as the town of Manhattan formed, and by August Palo Alto had been abandoned. Prospectors discovered a new ledge that assayed as high as $10,000 a ton, focusing even

more interest on Manhattan. The turning point for the booming camp came when mining promoter Humboldt Gates, known for investing only when success was practically guaranteed, began investing heavily in some of the new claims in the area.

Property speculation went wild during the boom. Lots sold for as much as $1,900. Soon the gulch was filled with saloons, hotels, assay offices, and a few schools. Telegraph and telephone service was brought in, and an electric power substation was built. A post office, with Ralph Stevens as postmaster, opened on December 25, 1905. At that time, the town already had three banks, a Wells Fargo express office, seventy-five frame buildings, and a population of 1,000. Two newspapers started circulation in January 1906. The *Manhattan Mail* was the first, making its appearance on January 10. Haworth and Anderson published it every Wednesday. The paper featured many photographs of the town and mines, which was rare for Nevada newspapers. By 1909 Haworth and Anderson had leased the *Mail* to Frank Garside. He left in 1910 to start another paper, the *Manhattan Post*. The *Mail* stopped circulation for three weeks before another publisher was found. The replacement, Roy Mighels, was replaced shortly afterwards, as well. A number of other publishers soon followed. The right match was never made, and the paper folded on June 24, 1911. The second newspaper in Manhattan was the *Manhattan News,* which started publication in January 1906 and folded on July 7, 1907.

Three other newspapers served Manhattan during the early part of the century. The *Manhattan Times* started publication on July 6, 1907, and folded on December 7 of the same year. Former *Mail* publisher Frank Garside started the *Manhattan Post* on October 25, 1910. After the *Manhattan Mail* folded in 1911 the *Post* was the only paper left in town. It continued publication until May 30, 1914, and went out with the following statement: "At the present time business conditions do not warrant the further publication of a newspaper." The last paper to serve Manhattan was the *Manhattan Magnet,* which William Godwaldt started on March 23, 1917. The paper, whose nickname was Queen of the Toquimas, enjoyed five years of publication. But the collateral backing the paper failed, and the *Magnet* was forced to fold on September 30, 1922. From then on Manhattan had to rely on Tonopah for newspapers.

Panorama view of Manhattan on February 18, 1906. (Nevada Historical Society)

Preserving the Glory Days

Two dramatic events took place in Manhattan in April 1906. The first was the shocking murder of Manhattan sheriff Thomas Logan. Exact circumstances of the murder are not known. The killer was Walter Barieu, described as a low-down gambler and a fiend. The setting for the murder was the Jewel Saloon, where Barieu and another gambler got into a wild fight. Sheriff Logan was summoned and was able to stop the conflict. But as Logan was leaving, Barieu pulled out his gun and shot Logan five times. Only the fifth shot missed.

Another account, which the *DeLamar Lode* printed on April 17, said that Barieu had hit a woman and that Logan had thrown Barieu out of the saloon. Barieu then shot at Logan through a side window in the saloon, but he missed. Logan ran out of the establishment, and then Barieu fired the four shots, mortally wounding the sheriff. In spite of his injuries, Logan supposedly disarmed Barieu and knocked him unconscious. There are other versions of the event, probably none of which are completely accurate. Sheriff Logan was extremely popular, and his friends organized the largest funeral Nevada had witnessed up to that time. It was held in the Odd Fellows and Eagle lodges at Tonopah. The case was well publicized and eventually resulted in Barieu's acquittal. The case showcased two future Nevada political giants. Key Pittman was the prosecutor, and Pat McCarran served as defense counsel.

Manhattan had only one week to recover from the Logan tragedy before another disaster struck. The San Francisco earthquake of April 1906 jolted Manhattan almost as much as it did San Francisco. Much of the mining activity in the Manhattan district was backed by San Francisco financiers, who withdrew their support to help rebuild San Francisco. Manhattan's banks closed, and many potentially successful mining companies also had to close. The population quickly dropped to a few hundred. Despite this setback, there were still many businesses in Manhattan in the summer of 1906. They included the Butte Cafe, the Giffen Mercantile Company, the Manhattan-Tonopah Brokerage Company, the Manhattan Lumber Company, the Nye and Ormsby Bank, the Bank of Manhattan, the Manhattan Transfer Company, the Manhattan Liquor Company, the Palace Hotel, the Tonopah Club, the Madden Saloon, the Manhattan Chop House, the Richard Hardware Company, the Manhattan Undertaking and Construction Company, the Nevada Rapid Transit, and the State Bank and Trust Company. Many mining companies were also active, including the Nevada Manhattan, the Manhattan National, the Manhattan-Virginia, the Manhattan Catbird, the Manhattan Eastern, the Jordan and McClellan, the Mayre, the Manhattan Gold King, the Manhattan Gold Ledge, the Manhattan American Gold, the Manhattan Century, the Bullfrog Rush, the Manhattan Venture, the Manhattan Joker, the Manhattan Reliance, the Manhattan Calumet, the Manhattan American Flag, the Manhattan Golden Crust, the Manhattan United States, the Manhattan Red Top, the Comet Gold, and the Manhattan Mizpah. New discoveries in September

1906 and June 1907 barely kept the town alive. The Manhattan Reducing and Refining Company completed the first mill in January 1907. The Rosario Mining and Milling Company and the Nevada Ore Reducing Company finished two others in early 1908.

The town's perseverance paid off when, in 1909, rich placer deposits were discovered on the edge of Big Smoky Valley, a few miles below Manhattan. The discovery revitalized the town to a certain extent. That year the town also received the first wireless telegraph in Nevada. The year was one of ups and downs for Manhattan. In January the State Bank and Trust and the Nye and Ormsby Bank both folded. In March a fire, which started in the Nevada Hotel, destroyed eight downtown buildings. As a result, the Manhattan Volunteer Fire Department formed. Electricity arrived from a substation at Millers in Esmeralda County, and during the line's construction the workers set a world record for line running. But the excitement was tempered in July when a gas explosion ripped through the Tyke mine, killing Albert Elton, Edward Hopf, and Roy Parr.

The town received a boost in February 1910 when the War Eagle mill started operations. Manhattan's mines yielded more than $616,000 worth of ore in 1910, and many prosperous years followed. The level of activity increased again in 1912 when prospectors discovered a rich new lode at the bottom of the already rich White Caps mine, located only a mile away from the town. A seventy-five-ton mill began operating at White Caps, and in April the Associated mill started operating in Manhattan. The 100-ton Big Four mill started up in March 1913. In town a school, which students attended until 1955, opened. The population of Manhattan rose to almost 1,000 during the next two years. After the 1920s production declined, and most operations shut down.

During Manhattan's lengthy production period many different mining companies were active in the district. The most important of these was the White Caps Mining Company. (See the section on White Caps for a detailed discussion of this company.) The Manhattan Consolidated Mine Development Company was another major force in the district. It was based in Tonopah and had mine offices in Manhattan. The company, which was incorporated in 1913 with $1 million in capital, issued 1,350,000 shares of stock at $1 a share and was listed on the San Francisco stock exchange. J. H. Miller was president, and M. N. Page was mine superintendent. Manhattan Consolidated engaged in extensive litigation with the White Caps Mining Company. The dispute wasn't resolved until 1917. The company's property consisted of five claims that covered about eighty-two acres. There was a shaft more than 600 feet deep that contained over 4,000 feet of workings. Ore removed from the property assayed from $18 to $30 a ton.

The Union Amalgamated Mining Company—which formed as the result

of a merger between Manhattan Amalgamated, Litigation Hill Merger Company, and Manhattan Earl—was another major Manhattan company. Union Amalgamated was based in Manhattan. C. F. Wittenburg was its president. The company owned six claims on Litigation Hill that produced more than $200,000 worth of gold and silver. The ore ran from $15 to $25 a ton. In addition to the six claims, the company also owned the Manhattan Milling and Ore Company's ten-stamp mill. Union Amalgamated reorganized in November 1917, and Wittenburg remained president. The company's new name was Manhattan Union Amalgamated Mines Syndicate. The change did not do much to improve the company's situation. The mill closed in 1918, and in July 1919 the company folded.

Two other mining companies were important in Manhattan's mineral production. The Manhattan Big Four Mining Company incorporated early in 1906 and had sixty claims in the district. The company dug a 500-foot shaft, but the high-grade ore soon ran out, and extremely low-grade ore replaced it. The company owned a 100-ton mill that closed in 1913 and then reopened briefly in 1917. The other company, which had originally been the Red Top Mining Company before it was reorganized in October 1912, owned sixty acres of claims in the district, including a 200-foot shaft. But it had to shut down in March 1918 because of heavy water seepage into the shaft.

Numerous smaller companies also worked the district. Some of these, like the Original Mining Company, lasted for many years but never hit the big ore bodies. Others, like the Zanzibar Mining Company, the Mammoth Gold Mining Company, and the Manhattan Dexter Mining Company were backed by wealthy businessmen who could bankroll extensive operations. But the value of the ore was so low that the companies folded.

Some Manhattan companies formed without first determining whether trying to recover the ore was feasible, and they succumbed before they were even established. Examples of such companies included the Manhattan Mustang Mining Company, the Manhattan Sunrise Mining Company, the Manhattan Copper Mining and Milling Company, and the Wolftone Extension Mining Company. The 1920s were slow for Manhattan mining, and two devastating fires destroyed most of downtown. On December 23, 1920, a fire that started in the Pine Tree Garage and Bank Saloon also burned the Manhattan Commercial Store, the Smith Tin Shop, the Connor Store, the Red Front Saloon, the Herd Store, the North Ferguson Drug Store, the Nevada Telegraph and Telephone Company, and twenty other buildings. Another fire in May 1922 burned the south side of the business district, including the Central, the Palace, the Merchants, and the Lloyd Hotels. The post office burned and was relocated to the Victoria Hotel. One week later another fire burned the Victoria Hotel and more of the business district. In addition, the two May fires burned almost fifty homes. Only some of the businesses and homes were ever rebuilt. In the 1930s the Manhattan Reliance Mining Company, which produced $500,000 worth of ore from 1932 to 1935, started a new revival in Manhattan. During this period three mills were running in the town. They were the Manhattan Consolidated, the War Eagle, and the White Caps. The Manhattan Placer Company was also active and by 1936 was producing $12,000 worth of ore a month. The Manhattan Gold Dredging Company bought its property in 1938.

In May the company built a 3,000-ton dredge in lower Manhattan Gulch. It constructed an artificial pond with water piped in from Peavine Creek. The operation recovered almost $4.6 million on a huge volume of processed ore: one cubic yard of ore only yielded twenty-one cents worth of pure mineral. The total value of ore mined in the Manhattan district is well over $12 million.

After dredging operations ceased in 1947 Manhattan began once more to move towards ghost town status. In the 1980s there was some new mining activity near Manhattan. The open-pit operation involved cyanide leaching. In the summer of 1979 the operation was moving into high gear. Argus Resources had also started deep mining operations in the spring of 1980 and reopened a 650-foot shaft on Litigation Hill. The shaft was filled with water up to the 400-foot level, but the Sierra Pacific Power Company brought power to the mine site so that it could be pumped dry. The company also controls the White Caps mine. Rich ore may flow again from the veins of the Manhattan mines. The town still has a population of fifty, and the post office and saloons remain open. The old concrete electric power substation, which previously served as a curio shop for Manhattan area souvenirs, was dismantled in 1985 during recent mining operations. Gasoline, some groceries, and a public telephone are all available in Manhattan.

The remains at Manhattan are extensive and very interesting. In addition to numerous cabins and small houses, the old stone post office and a number of false-front buildings still stand. The Catholic Church in Manhattan, which was moved from Belmont in 1908, was restored in the 1970s. A visit to the Manhattan Cemetery, located half a mile west of town, is a must.

McCann Station

(Swastika)

DIRECTIONS: *From East Belmont, head south for 6 miles. Head left for 1 mile. At the fork, take the left road and follow it for 6 miles to McCann Station.*

McCann Station was a stop on the Belmont-Tybo-Eureka stage route beginning in the 1870s. Barney and Grace McCann owned the station and ranch. Barney was Bernard McCann's son, and Grace was his widow. All three were Irish immigrants. The station was the only stop between Tybo and Belmont. The McCanns provided lodging and meals to weary travelers.

While the road saw little use after the 1880s, the McCanns, along with Bernard's brothers, John and William, continued to run the ranch. William was a prominent miner and prospector throughout Nye County. Bernard had served many years as a county commissioner. He died in 1879 while he and Grace were running boardinghouses in Tybo and Belmont. The couple had seven children, most of whom grew up at McCann Station. In 1920, when Barney was running the ranch, he was murdered by Robert Hughes while in Belmont. Grace died in 1925, and the ranch was eventually abandoned. It was still occupied in 1930, but since then it has virtually disappeared. Only wood scraps and a foundation mark the site.

Some limited mining activity took place in the canyon beginning in 1912. In May Jim Hughes, Aaron Castle, and Harry Stimler found a vein of rich silver ore and organized the Swastika Mining District. George Wingfield took a bond on the property, and a small camp began to form, but after he gave up interest quickly faded, and everyone was gone by the end of the year. Nothing remains at the camp.

Millett

DIRECTIONS: *From Round Mountain, head west on Nevada 378 for 2.7 miles. Exit right (north) on Nevada 376 and follow it for 21½ miles to Millett Ranch, located on the east side of the highway.*

In 1873 Charles Scheel built a small ranch on the eastern slope of the Toiyabe Mountains. He ran the ranch until 1896, when Albion Bradbury Millett bought the property and gave it his name. Millett had owned the Twin River Ranch, but after marrying Scheel's widow, Christina, he moved to her ranch. Millett was very active in local mining and had served as a state senator.

Millett remained just a ranch until 1905, when prospectors flooded the North Twin River Mining District after major strikes were made in Round Mountain and Manhattan. In early 1906 prospectors discovered rich ore in Park Canyon, just west of Millett. Soon a small townsite had been platted next to the ranch. By June 1906 Millett and the surrounding camps in the district were home to more than 300 gold-hungry residents. On May 3, 1906, a post office, with Albion Millett as postmaster, was established to serve the entire North Twin River Mining District. Christina Millett opened a large general merchandise store in 1906. The impressive Lakeview Hotel, a two-story structure noted for its comfortable rooms, was also built during the 1906

Overview of Millett in 1909. (Frances Humphrey collection, Central Nevada Historical Society)

Preserving the Glory Days

rush. Soon a number of saloons, a blacksmith shop, a wagon-repair shop, and other buildings were also erected. The Smoky Valley Hotel, which G.H. Brown owned, was completed in 1909.

Most of the ore mined in the district was shipped through Millett to Austin, and soon Millett became a very important shipping center for the Toiyabe and Toquima range mines. Unfortunately, however, most of the mines faded quickly, and by 1911 Millett's population was down to fifty. By 1916 the mining district was dead, and in November 1923 the Milletts sold their ranch to Will and Archie Farrington for $45,000. The Millett post office, located in the general store, stayed open until July 3, 1930.

The remains of Millett are now on the George Frawley Ranch. Dugouts mark the sites of the many buildings. The original Scheel ranch house also still exists, along with the crumbling stone walls of the Millett general store. Be sure to ask permission before exploring.

Minnimum's

(Bauman's)

DIRECTIONS: *Minnimum's is located ¼ mile east of Nevada 376, 34 miles north of Round Mountain.*

Minnimum's station formed in early 1862 as part of the Belmont–Austin stage line and was on A.E. Minnimum's ranch. It became more important during the 1880s and 1890s when mining activity in the Toiyabe Mountains increased. Some mining occurred directly west of the station, and

A few stone cabins in good shape still stand at Minnimum's. (Shawn Hall collection, Nevada State Museum)

a number of prospectors resided in Minnimum's while working claims in the mountains. Minnimum died in 1889, and his wife immediately sold the ranch and left the area. The station never really boomed, but its peak population was more than twenty-five—not bad for a stage stop. After the turn of the century the station began to become obsolete, and it was abandoned in the early teens.

The remains at Minnimum's are beautiful. A few old stone cabins with brush and mud roofs still stand at the site. Stone foundations are scattered near the cabins. The site is a definite must, not because of the number of its ruins but because of their quality.

Monarch

DIRECTIONS: From Tonopah, head east on US 6 for 5.5 miles. Exit left on Nevada 376 and follow for 13.3 miles. Exit right and follow old Nevada 82 for 25 miles. Exit right and follow this road southeast for 4 miles to Monarch.

Monarch resulted from one of the biggest promotional scams in Nye County history. The town was the brainchild of Reverend Benjamin Blanchard, who platted the townsite during the summer of 1906. He also had ranch estates and mining claims platted. Blanchard began a tremendous promotional campaign, including advertisements in Tonopah and Virginia City newspapers, and personally travelled all over the western part of the country. He ended up selling more than 2,300 lots to the unsuspecting public. By August, King Ryan had started the Manhattan-Monarch stage line. A number of mining companies were formed, including the Beroni Copper Company, Continental Consolidated Mining Company, Howell Mining Company, and National Gold and Copper Company, all of which had Blanchard as manager. He also ran the Colonial Investment Company and the Monarch Investment Company. A paper, the *Monarch Tribune,* began publication on August 18, 1906, with Lester Haworth as editor and manager. During its one-month existence, it acted as a propaganda medium for Blanchard. Additional freight and stage lines were set up to the town, and the telephone and telegraph company began to put up a service line. The Western Pacific Railroad even entertained thoughts of having a spur line to Monarch.

By October 1906, the townsite had a population of 150 and contained a number of stores and shops along with three boardinghouses and two hotels. Almost all of these buildings were transported to Monarch from other camps in Nye County. An official order to establish a post office was given on October 15 (William Williams, postmaster), but soon thereafter, Monarch's bubble burst.

By the end of October, Blanchard had sold 2,400 town and ranch lots.

After collecting more than $75,000 from the trusting people, he left town with the money and also left behind more than $73,000 in debts. Blanchard left under the guise of going back East to ascertain why the eastern financial backers had not sent funds. Once out of town, Blanchard disappeared with all of the money. Within weeks, Monarch was also well on its way to disappearing. By November 1906, only a handful of people were left, most acting as watchmen for the empty buildings. The post office was rescinded on March 25, 1907, putting the final nail in Monarch's coffin. The ruins at Monarch are extremely scarce. Only a few collapsed outhouses and some scattered wood boards mark the site.

Moores Creek Station

DIRECTIONS: *Moores Creek Station is located 5 miles north of Jefferson.*

Moores Creek Station served as an occasional stop on the Belmont–Austin road during the 1870s. Moores Creek was an alternate route for the stage when the normal route was impassable. It was here that in December 1878 John Jett, discoverer of the Jett Canyon mines, died while traveling. By the 1880s the station had been abandoned, and a small ranch began operations. The creek was named for John A. Moore, who settled there in the 1880s. Moore was extremely poor until he leased a mine at Morey and made $30,000. He used the money to start the ranch. In 1904 a local rancher, C.W. Anderson, discovered some gold and silver ore. Not much happened in the area until 1908, when a small rush led to the formation of a little camp. By April a canvas camp of about twenty had materialized. William Trembath, T.L. Sullivan, C.M. Dull, and Pat Murphy leased Anderson's Baby mine, removing ore worth $40 a ton. The excitement died down by the end of the summer, although miners made some additional attempts to extract ore during the next few years. Only a couple of collapsed shafts mark the site today.

North Manhattan

DIRECTIONS: *From Manhattan, head west on Nevada 377 for 3½ miles. Exit right and follow the road for 3½ miles. Exit right and follow this road for 2 miles to North Manhattan.*

North Manhattan, an extremely short-lived mining camp, formed in early 1905 and was abandoned late that summer. The camp was high up in Bald Mountain Canyon near some small gold veins. A tent camp of twenty quickly materialized, but the veins played out just as quickly, and the canyon was soon abandoned. In 1917 the Putney Gold Mining Company began working the area, and in March 1918 it started a small five-stamp mill, but the mill only operated for a few months. In 1920 J. D. Johnston worked the Newport Group and planned a fifty-ton mill, but the ore values dropped before it was built. The last bout of activity at North Manhattan took place in 1930, when Charles Lohman worked the Anaconda claim group but produced little. North Manhattan has seen no additional activity. Only small tailing piles mark the site.

Northumberland
(Bartell) (Monitor) (Revive)

DIRECTIONS: *From Belmont, continue north on old Nevada 82 for 24 miles. Exit left and follow the road for 6 miles. Exit right and follow the road for 1½ miles to Northumberland.*

It appears that two separate groups of prospectors discovered rich silver ore on the eastern slope of the Toquimas in July 1866. According to the *Mountain Champion,* Bob Logan, Harry Fletcher, Buckhorn, Kincaid, and Charnock discovered the Northumberland mine. The *Nye County News* credited D. B. Thrift, J. C. Merrill, and W. N. Cummings with the discovery of the Lady Cummings mine. The second group organized the Detroit Mining Company. Its active mines were the Lady Cummings, the Fowler, the Clark, the Silver Bar, the Mountain Top, and the Silver Mount. The Northumberland Mining District, which was named for the English county north of the Humber River, organized in 1866. It is possible that Logan was from that part of England. A little camp formed in October 1868 when a company moved a small, ten-stamp mill to the Northumberland area from Indian Springs. The Quintero Company ran the mill. The ore did not last, however, and by 1870 the district had been abandoned. The mill was at Learville, half a mile away. (See the section on Learville for a more detailed discussion of the mill.)

In 1875 the discovery of two new mines, the Monitor and the Blue Bell, revived the camp, whose name had been changed from Northumberland to Monitor. The camp was also known as Bartell for a while before its name was changed back to Northumberland in 1880. Many printed accounts, including the first edition of this book, incorrectly identify the town as Bartlett. It was actually Bartell and had been named for Peter Bartell, president of the Bartell Silver Mining Company. The townspeople selected the name in October 1879, and it replaced Revive as the name of the town. By that time Bartell's population stood at about fifty, and the town included an unofficial post office, a store, a boardinghouse, and a number of saloons. A freight line to Austin was set up, and shipping ran $20 per ton. In August the Belmont–

The large Northumberland Mill during its peak production period during the 1930s. The camp was to the left, just out of view in the photo. Large foundations remained in 1980 but have disappeared under a huge tailing pile from current mining operations. (Special collections, Getchell Library, University of Nevada)

Foundations of the Northumberland Mill in 1980. The foundations are now buried under tons of tailings. (Shawn Hall collection, Nevada State Museum)

Austin stage route began running through Northumberland Canyon rather than going through Smoky Valley to Monitor Valley, making an overnight stop at the town. In November 1879 a new ten-stamp mill started up, but it only functioned for three months before closing. The mill was the old Metacome mill (in Lander County). S. Slusher and Adam Stoneberger, who were partners in Bartell's company, had purchased it and moved it to Northumberland.

Activity in the district dwindled once more, and by 1881 Northumberland was again abandoned. The Bartell company went bankrupt, and the property was sold in May 1880 at a sheriff's auction. Bartell then returned to Belmont, where he worked his Courthouse mine.

Another revival took place in Northumberland in 1885 and 1886. Sixty people began working the district. An official post office, with John McPherson as postmaster, opened on February 24, 1885, but mining activity quickly dwindled, and the post office closed on July 16, 1886. McPherson owned the mines and reopened the mill for a while. Things remained uneventful in the district until 1908, when new silver discoveries prompted a spurt of activity. From 1887 to 1893 Adams, Brewer and Company worked the mines but met with only minimal success. The Reno Goldfield Mining Company built a new 100-ton cyanide and concentration mill, which remained in partial operation until 1917. The district became fairly quiet, with only leasers periodically working the area. The Northumberland Mining Company bought the mill and a number of claims in 1939 and began intensive operations. With machinery brought in from Weepah, the company enlarged the mill to a capacity of 325 tons, and it built an extensive camp to house its seventy-five workers. A school opened in the town, and a voting precinct was established. The miners also formed a softball team, the Golddiggers, which played in the Nevada Central Softball League. By 1941 the open-pit mine was yielding 10,000 tons a month, but in October 1942, due to a War Production Board order, the mine was forced to close, and all operations were suspended. The company produced well over $1 million worth of both silver and gold before closing. No activity occurred in Northumberland after 1942 until the early 1980s. A barite mine was active for a number of years at the mouth of the canyon, but it has since ceased to operate.

Fifteen years ago Northumberland's remains were fairly extensive. A number of wood buildings, which were left from the Northumberland Mining Company's activities, still stood. There were a few stone ruins dating back to earlier bouts of activity. The most impressive ruin was the huge, 325-ton mill, with its enormous concrete foundations. As recently as the 1940s this had been an operating mill, but the harsh elements had reduced the once-proud structure to nothing more than a pile of rubble. Also interesting were the wood supports of a tramway that ran from the mill to a flat meadow lower in the canyon, near the old Learville townsite. These supports could be seen for one and a half miles on the side of the canyon, where the tram had moved the large volume of tailings away from the mill so that they would not clog up the narrow canyon.

Unfortunately for historical Northumberland, in 1983 the Cyprus Mines Corporation begun an open-pit heap leach facility. In recent years the pit and leach pads have completely obliterated the townsite. This is a large operation that in 1988 produced 130,000 ounces of silver and 30,000 ounces of

Overview of the camp of Northumberland in 1980. (Shawn Hall collection, Nevada State Museum)

gold. By 1990 Western States Mineral Corporation was running the mine. However, difficulties in obtaining permits to expand it led to the suspension of operations in 1990. It appears that mining will resume soon. The Northumberland Cave is on top of the hill behind the pile. At the mouth of Northumberland Canyon are the ruins of an old chlorination mill that was used from 1870 to 1880. Nearby is the open pit of a barite operation that was active in the late 1970s and early 1980s. Many foundations of that small camp remain. The drive through Northumberland Canyon is extremely scenic. When exploring, be on guard for rattlesnakes.

Pine Creek

DIRECTIONS: *From Belmont, continue north on old Nevada 82 for 15 miles to Pine Creek.*

Pine Creek was a stage stop on the Belmont–Austin stage line from the 1860s to the early 1880s. During this time, a small ranch and a few other buildings existed at the site. The stage station was a stone structure, as were most of the ranch buildings. The area never had a population of more than twenty-five. A post office was active at Pine Creek from May 9, 1873, to January 4, 1875. During this time, Elias Erickson was postmaster. It was active again from September 29, 1879, to January 18, 1881. This time Allen Crabtree was postmaster.

Although the Belmont–Austin stage route was rerouted through Jefferson, Pine Creek has been maintained as a ranch ever since it was first established. For many years George Ernst, who served as county surveyor and was involved in mining business throughout Nye County, owned the ranch. The original stage building and ranch headquarters remained standing until the early 1970s, when it was destroyed by fire. Owners built another two-story building at Pine Creek early in this century, and it served as a boardinghouse and a summer home. This building stood just to the left of the present ranch house. Tasker Oddie purchased Pine Creek from Ernst in 1902 for $9,000 and remained its owner while he was Nevada's governor. He used the large wooden house as a summer home for a number of years before selling the ranch to Harry Stimler and William Marsh for $200,000 in December 1908. The price also included the Corcoran, the Haystack, the Cook, the Northumberland, the Punch Bowl, and the McGonigal ranches. In September 1911 the Nye County Land and Livestock Company took over the ranch. The Tonopah Banking Corporation, which brought Chinese pheasants in, ran the company. Later, in 1917, Pine Creek became part of the large United Cattle and Packing Company. O.K. Reed, of Kawich fame, also owned the ranch when he controlled most of central Nevada's cattle industry. During the spring of 1936 CCC workers built the Pine Creek campground. The United Cattle Company sold the ranch in 1940 to John Wardlaw, who in turn sold it to Harold and Warren Hunt in 1948.

Today only one of the original buildings remains at Pine Creek. It is currently used for cold storage. The Hage family presently owns the ranch. Nearby Pine Creek Campground is one of the more picturesque campgrounds in Nevada and offers hiking and great fishing opportunities.

Potts

DIRECTIONS: *From Belmont, continue north on old Nevada 82 for 37 miles. Exit right and follow the road for 1½ miles. Exit right and follow this road for ¾ mile to Potts.*

Potts, a small but prominent ranching settlement, was named for William Potts, who first organized the ranch in the 1870s. The Potts Ranch earned a reputation as one of the best cattle operations in the state. The government granted it a post office and then rescinded the grant in 1893, but the office opened officially on August 12, 1898. William Potts served as postmaster. The post office continued to operate until October 31, 1941. All the postmasters—Bessie, Mamie, Maude, Irene, and Anna—were members of the Potts family. For a while the Potts Ranch was the site of a ranger station. At its peak, the Potts family also ran the Wilson, the Morgan Creek, and the Butler Ranches. In 1944, not long after the death of George Potts, son of William, in 1942, the family sold all the ranches to Harvey Sewell and O.G. Bates. The Wilson Ranch became Potts's headquarters, and the Potts Ranch saw little use.

The Potts ranch house as it appears today. (Shawn Hall collection, Nevada State Museum)

The ranch complex was abandoned many years ago, but the remains are very interesting. The ranch house still stands and is in fairly good condition. A number of stone dugouts and two other buildings are also left. An old corral is just to the west of the buildings. A rare old slaughter wheel that stood at the corral in the late 1970s has since disappeared. A number of other ranches in the area still use the site. Potts is well worth the trip. When visiting Potts, don't miss Diana's Punch Bowl, a nearby hot spring located just south of Potts.

Round Mountain

(Gordon) (Shoshone) (Brooklyn)

DIRECTIONS: *Round Mountain is located 2.7 miles east of the Junction of Nevada 378 and Nevada 376, 54 miles north of Tonopah.*

Prospectors discovered gold at Round Mountain as early as 1905, but it was not until February 1906 that more substantial discoveries aroused sincere interest in the area. In February three men, including Louis Gordon, discovered high-grade gold ore on a hill known as Round Mountain. A small camp formed at the base of the mountain and was named after Gordon. By May Bagley and Long had organized the Round Mountain Townsite Company, and three stage lines were running. The Gordon post office, with Chester Olive as postmaster, opened on June 28, 1906, and soon the camp's population was close to 400. A weekly newspaper, the *Round Mountain Nugget,* began publication on June 2, 1906, and lasted until 1910. The paper greeted residents of Round Mountain with the following message:

With this issue a new enterprise is launched, a new voice is raised to tell of the greatness of the state of Nevada and the wonderful wealth that nature has stored in her hills. And what is more in accordance with the eternal fitness of things than that voice should rise from a new camp, the youngest of the sturdy brood that is growing up as never children grew before to make great the name of the grand old site.

When in days of old, when knights went forth to battle, it was the custom for each to bind to his helmet a glove, a piece of ribbon or some kindred token, and with the name of the lady who gave it upon his lips, go bravely on to meet what fate might have in store for him, high of heart and strong of arm in the knowledge of his lady's constancy.

Even so in this its initial appearance in the journalistic lists, does the Nugget take upon itself the cause of Round Mountain, the youngest lady of them all, and even as did the knights of olden days, does the Nugget defy the world to tramp the colors it bears in the dust.

On three things should every good American boast: of his country; his state and his town. Backed by its confidence in its country and its state, the Nugget makes its entrance into the world of newspapers, strong in its belief that its town, Round Mountain, a camp whose life is numbered by months, nay by weeks, will, ere many months have passed, be granted a place among the greatest camps of the state by the common consent of a couple who have learned to know its value from the wealth it has produced.

A number of major mining companies became active in the Round Mountain district during this period. On March 4, 1907, the name of the growing town was changed to Round Mountain. Two adjacent smaller townsites, Shoshone and Brooklyn, were laid out. The Shoshone site, located on the banks of Shoshone Creek, grew to include the Merchant's Bar, owned by Joe Mason and Jack Hanna; the Duke and Andrews restaurant; a lodging house; a feed yard; and a saloon. Free lots were offered. However, Round Mountain was growing so fast that plans changed, and the larger townsite of Round Mountain became the main town.

By 1907 there were daily stages running from Round Mountain to Tonopah. Fresh freight supplies arrived daily. The town contained many wood structures housing mercantiles, saloons, brokerage agencies, a school, and a substantial library. That same year Thomas Wilson discovered the Dry Wash placers just below the rich Round Mountain mines. Wilson, a resident of nearby Manhattan, consolidated his placer claims into the Round Mountain Hydraulic Mining Company in 1907. Water for the operation first came from the Jefferson and Shoshone Creeks. In 1915 the company completed a new and more efficient pipeline to Jett Canyon, across Smoky Valley. These placers accounted for a large percentage of Round Mountain's total production figures.

In 1906, soon after its discovery, two mining companies began operations in Round Mountain. Round Mountain Mining Company was the principal and also the first operator in the district. It was incorporated in March 1906. The company, of which Louis Gordon was president and general manager, had offices in Round Mountain and San Francisco. The company controlled almost 1,000 acres of mining claims, which ranged from gold to silver and from lode deposits to placers. The company built a 180-ton stamp mill near the Sunnyside mine, which was the largest mine in the district. Sunnyside had a 1,000-foot incline shaft and miles of lateral work. The ore values were low, averaging about $7 per ton, but mining was still profitable because the mill could treat more than 1,100 tons of low-grade ore daily. In 1914 the company bought out the Round Mountain Sphinx Mining Company, which controlled the Los Gazabo claim. This claim proved to be a valuable addition. The value of the ore produced by the company's mines during their first ten years (1906 to 1916) was a little more than $3 million. The Sunnyside mine and mill continued to operate until 1921.

The second mining company to come to Round Mountain was the Fairview Round Mountain Mines Company, which was incorporated on June 25, 1906. Louis Gordon was president of this company, too. Fairview controlled twelve claims in the district, including major mines named the Daisy and the Fairview. A six-stamp amalgamating mill built next to the Daisy mine had a daily capacity of about fifty tons of ore.

By the beginning of 1909 there were six mills operating in the district. Businesses in Round Mountain included the Tarbell Hotel; the Chester Olive general store; the Round Mountain Bank, owned by Chester Olive; the Round Mountain Mercantile Company; the Ideal Restaurant and Lodging House; the Millet Supply Store; the Miles Grocery; the Walker Saloon; the Overland Hotel; the Round Mountain Athletic Club; the Miners Union Hall; the Round Mountain Water Company; and the Round Mountain Stock Exchange. The town also had a school, a library, and a hospital. The local baseball team

provided weekend entertainment for the residents. Power lines to the town were completed in May. The hydraulic operations were Round Mountain's main source of support. These operations led to the formation of two small camps near the mines: Shoshone (on Shoshone Creek), and Brooklyn (near the mouth of Jefferson Canyon).

It was not until 1919 that other large mining companies came to the district. Fairview Extension Mining Company was incorporated in July 1919 and was controlled by Fairview Round Mountain Company interests. Louis Gordon was also president of this company, which controlled three major claim groups: the Alta, the Shot, and the Blue Jacket, the latter of which it purchased from the defunct Round Mountain Blue Jacket Mining Company. The company mainly conducted surface operations. The Blue Jacket mine was only about 100 feet deep, but in total the Blue Jacket group produced almost $100,000 worth of ore before shutting down in the early 1920s.

The Nevada Gold Development Company was the next major company to move to the Round Mountain district. Incorporated in July 1925, the company took control of the Tom Wilson placers. A shaft dug on one of the company's other claims eventually reached a depth of 300 feet, with lateral workings of 400 feet. Ore from the mine averaged only $2 a ton. The company was absorbed into the Nevada Porphyry Gold Mines Company in 1928.

The Round Mountain Mining Company was reorganized in 1926 and was renamed the Round Mountain Mines Company. The company enlarged the old Sunnyside mill to 200-ton amalgamation size. The mill also contained twenty stamps, a small Huntington mill, and a few smaller tube mills. The company continued to conduct hydraulic operations, mostly in the spring, when there was an abundance of water. It spent more than $200,000 on the Jett Canyon water pipeline to Round Mountain. The pipeline enabled the placers to produce for a few more years. Over the next two years the company gained control of the district's other companies.

In 1928 the company gave its option to the newly formed Nevada Porphyry Company. This company consolidated all the mining activity in the Round Mountain district and even today has interest in local mining properties. In 1935 underground mining was deemed too dangerous, and the company folded. By 1939 the town's population was still 234. The value of total production through 1940 was $7.8 million. The Round Mountain Gold Dredging Company built a 17,000-ton placer mill in 1949. Hopes were high that placer mining would once again prove profitable, but the ore values became extremely poor, and operations ceased in 1952. Mining resumed in 1955, but the mill closed in 1959 after producing $3 million worth of ore. Round Mountain was quite ghostly until the early 1970s, when life came back to the little town. By then the population had shrunk to about 100.

The Copper Range Company began extensive new surface operations on the side of Round Mountain in 1970. The Smoky Valley Mining Company started an open pit in 1976, and the first gold bar was poured in February 1977. In 1985 the company sold the operation to three partner companies: Echo Bay, Homestake, and Barigold, and the Round Mountain Gold Corporation formed.

By 1989 the mine was producing 300,000 ounces a year, and that yearly production figure has continued to grow. The din of heavy machinery is ever present, as the area is still being worked around the clock. Microscopic gold is the mined product, and huge tailing piles that extend for miles illustrate the size of the operation. An unexpected occurrence recently caused the company some problems. Miners hit a rich gold quartz vein, but the mill wasn't designed to treat true gold ore. After some adjustments were made, the new ore was successfully treated.

Most of the town's namesake, Round Mountain, has been carved away as the open pit continually expands. The expansion is troublesome in certain ways. Parts of the main ore body are under the town of Round Mountain, and eventually the town will be either moved or razed so that mining can continue. Present-day mining has led to the formation of a new town called Hadley in the valley below Round Mountain. Hadley, named after an officer of the company, has become larger than Round Mountain is today. From 1977 through 1993 the mining operation produced 1.6 million ounces of gold and 24.1 million ounces of silver.

There is still much to see in and around Round Mountain. On the hill just behind the town are two mills, both very dilapidated. A large number of abandoned mines are scattered throughout the town. In Round Mountain, many buildings from the early days remain. Plan to spend a day wandering through the picturesque town. Before the current mining activity began, total production figures for the Round Mountain mines were well in excess of $10 million. The mines are now producing easily more than a million dollars worth of ore every year.

Rye Patch

DIRECTIONS: From Tonopah, take US 6 east for 5½ miles. Then take Nevada 376 north for 11 miles to Rye Patch, located on the east side of the highway.

Rye Patch, in addition to providing water for the Hannapah Mining District and for the town of Tonopah, was also a ranching and mining area. The Sam Jack mining group was located on a ridge just east of the Rye Patch pumping station. This group consisted of fourteen claims owned by Earl Mayfield and Jack Clark. The group dug a number of shallow shafts, but the ore assayed at only $5 per ton, and the claims were never developed to any great extent. A number of small ranches in the Rye Patch Wash are the sites of the only activity still taking place in the area. At the pumping station itself a small recreational area has been created, and residents of Tonopah are frequent visitors.

Silver Glance

DIRECTIONS: From Hannapah, continue northwest for ½ mile to Silver Glance.

Silver Glance was Hannapah's sister camp. The mainstay of the camp was the Silver Glance mine, which prospectors discovered in 1905. The townsite was set up in 1906, and the real estate promoters who had worked Hannapah also promoted Silver Glance. They were even less successful there than they were at Hannapah. The closing of the Silver Glance mine in 1909 spelled the end for the camp, although revivals did take place in later years.

In 1919 the Silverzone Mines Company, which William Gray and J.S. Clark ran, reopened and began working the Silver Glance mine. They used the glory hole method, following a 125-foot-deep, 150-foot ore shoot. Gray and Clark leased out their holdings after 1920, and by 1921 the leasers had given up, too. The Silverzone Extension Mining Company was also active in the district in 1919. George Quigley, O.G. Walther, and J.W. Hanson—all of Tonopah—had organized the company in 1919. The company owned five claims just west of the Silver Glance mine. A 400-foot shaft sunk in 1919 and 1920 led to ore that assayed as high as $145 per ton. By 1922 the ore had faded, and the company folded late that year. The district was silent until 1928.

In 1928 the Silver Glance Mining Company was incorporated, and it purchased the old Silver Glance mine from Gray and Clark. Intensive exploration of the mine rewarded the new owners with a rich ore body that yielded almost $300,000 worth of silver in 1928. Unfortunately, the ore body was in a pocket, and by early 1929 the pocket was empty. When the company folded, so did Silver Glance. Mark Bradshaw took over the mine in February 1930 and put two shifts to work, but in April they struck heavy water, and work was suspended for good.

Only a few substantial buildings were ever built at the townsite, and there is no trace of them now. The only sign of what used to be Silver Glance is the Silver Glance mine glory hole. The townsite was platted just south of the mine, but it is impossible to locate today.

Spanish Spring

DIRECTIONS: *From Tonopah, head east on US 6 for 5½ miles. Then take Nevada 376 north for 13.3 miles. Exit right and take old Nevada 82 (Belmont Road) north for 12 miles. Exit left and follow this road for 3¼ miles to Spanish Spring.*

Spanish Spring was a little-known but important stage stop on the Pioneer stage line's Tonopah-Manhattan run. Established in 1905, the stage stop provided water to those about to make the tough climb over the mountains to booming Manhattan. The stage line was abandoned in August 1907 after the station was completely destroyed by fire.

A murder took place in Spanish Spring in March 1910. Taug Darrow killed William Baker, a rancher and Darrow's employer. Darrow was captured after a long search, and he spent the rest of his life in prison.

D. S. Llewellyn discovered tungsten at Spanish Spring in March 1916. He sold the claims to Theodore Bell for $100,000. Bell organized the Spanish Springs Mining Company and built a small mill, but the claims proved to be of little value, and when they were abandoned in 1917 they had produced nothing. One stone cabin and some scattered wood remains mark the site. The spring still flows, but very slowly, and it gathers in a stagnant pool. Do not plan on finding any drinking water at Spanish Spring.

Springfield Mining District

DIRECTIONS: The Springfield Mining District is located 8 miles south of Wilson's and 2 miles west of Diana's Punch Bowl in Ikes Canyon.

The Springfield Mining District formed on November 24, 1874, after E. Crane discovered the Sheba mine. As a result of his find, about a dozen prospectors came to the district. No one found much of value in the district in 1875, however, and the ore from the Sheba mine was worth little. The area was abandoned by 1876, and no other activity has taken place there since. Only the ruins of a small stone cabin are left in Ikes Canyon.

Stone House
(Smith's Station)

DIRECTIONS: From Belmont, continue north on old Nevada 82 for 7 miles. Exit left and follow the road for 1½ miles to Stone House.

For a number of years, Stone House was a stop on the Belmont–Austin stage line. L.D. Smith built the two-story stone station in the early 1860s, and weary travelers used it for the next thirty years. Many memorable dances took place in the ballroom on the second floor of the station, and they attracted people from as far away as Hot Creek.

Smith was killed on March 6, 1869, during a gun battle with E.L. Musick, Smith's employee. The dispute occurred over a team of oxen. Smith had purchased the oxen from Musick and was planning to resell them, which upset Musick. He followed Smith down the road from Stone House and killed him. For many years after Smith's death, the station was known as Widow Smith's place. M.W. Corcoran, in whose honor a nearby canyon was named, bought the ranch and station in the 1880s. After the stage stopped running in the mid-1890s, an Indian family named Hooper bought the station. They erected new buildings around the old station house and used the small complex as a ranch. Albert Hooper and his father, Tim, did a lot of prospecting in the hills west of the station. They staked a number of claims, but none turned out to be really prosperous. After his father died, Albert continued to run the ranch until the mid-1960s. The ranch has been abandoned since then, and the nearby Pine Creek Ranch now owns it.

Remains at Stone House are fairly extensive and interesting. The old station house is in relatively poor condition. There is no floor on the second level, and a wooden cabin has been built inside the station's shell. Although it looks strong from the outside, a quick look inside reveals large cracks, which indicate that the building is not going to last much longer. A portion of the roof recently collapsed. A number of newer wood cabins also remain near the

station house. Stone House is a definite must because of the station house's beauty. Spring water is available at the site.

Sumo

(St. Elmo)

DIRECTIONS: From Manhattan, backtrack to Nevada 376 and head south for 7 miles. Exit left and follow the road for 1½ miles to Sumo.

A short-lived camp named St. Elmo that sprang up and died during the early days of the Belmont boom produced nothing, and it wasn't until 1905 that serious mining took place. The new camp of Sumo, built by John Harris—manager of the Tonopah Northern Company—formed in early spring. By April the Tonopah Northern Mining Company, the Rambler-Tonopah Mining Company, the Willow Springs Mining Company, and the Tonopah Gold and Copper Mining Company were all active, and the camp had a population of forty. But the copper deposits were shallow, and this combined with depressed prices caused the camp to completely collapse by September. No one has exhibited any interest in the district since. Only shallow shafts and wood scraps show that Sumo ever existed.

Tate's Station

DIRECTIONS: From Round Mountain, head west on Nevada 378 for 2.7 miles. Exit right (north) on Nevada 376 and follow it for 19 miles to Tate's Station.

Tate's Station was an important stop on the Austin–Belmont stage in the 1880s and 1890s. Thomas Tate and his wife organized the station in 1886. Tate also had a mail delivery contract—which he fulfilled for more than thirty years—long before they established the station. He was influential in getting the first school built in Smoky Valley. Buildings at the station included the stone station house, a small lodging house, and a number of stables. The Tates closed the station in 1901 and sold out in 1907. They then moved to Big Pine, California. The Tates' daughter, Myrtle Tate Myles, became a well-known Nevada historian. Tate also provided part of Jim Butler's grubstake for the trip during which Butler discovered the riches at Tonopah. Eventually the site was incorporated into a ranch, which is still in operation today. The old station house and one of the sod stables remain.

Van Ness

DIRECTIONS: From Barcelona, backtrack for 1 mile and then exit right. Follow this extremely rough road for 2 miles to Van Ness.

Van Ness was a small mining camp located two miles from Barcelona. The Raymond Van Ness Mining Company owned and ran the camp. C. E. Van Ness, the company's president and manager, had his offices in Tonopah. The company had six claims in the district and began working them in the early 1920s. The claims consisted of cinnabar ore mixed with granite and shale. The ore was almost 5 percent mercury. The workings of the company included two shafts, both approximately 100 feet deep. The shafts followed a vein of ore varying in width from four feet to six feet. By 1929 the company had completed the development. It hired a full crew and established a camp at an altitude of 9,000 feet. The camp consisted of a couple of boardinghouses and a cookhouse, and it had a population of twenty. In 1929 Van Ness Mining built a forty-ton Gould rotary furnace to help smelt the mercury ore. That same year the company became active in the Ellendale district. It made its first shipments of quicksilver in June 1930, and by October the mines had produced $22,000 worth of the mercury. In December—the company's best month—it shipped $25,000 worth of quicksilver. It continued to produce well for a few years, but the cinnabar ore slowly ran out, and by 1935 the camp was abandoned. It wasn't until April 1940, when John Connolly and Homer Williams reopened the mines, that Van Ness became active again. The pair built a new boardinghouse and restarted the furnace, but the operation was short-lived, ending in 1941.

There are a number of interesting remains at Van Ness. The two mines retain most of their workings. Ruins of a number of cabins are scattered around the mines. Curious about a local rumor that an old Spanish fort was located just north of Van Ness, the author trudged through the area for a number of hours but was not able to find it. Perhaps a luckier explorer will one day come across this rare relic of early Nevada history—if it in fact exists. The road to Van Ness is very rough and has many washouts. It is passable only with a four-wheel-drive vehicle.

White Caps

DIRECTIONS: From Manhattan, head east for ¼ mile. Exit right onto a poor dirt road and follow it for 2½ miles to White Caps.

White Caps was a small mining center located a few miles from Manhattan. The Dexter Mining Company discovered and worked the mine, which the War Eagle Mining Company worked later. The White Caps Mining Company was incorporated in June 1915 with $200,000 in backing capital and was the first company to seriously work the area. It purchased two claims, which the Dexter White Caps Mining Company originally owned,

The White Caps Mine as it was in the 1920s, before the mine and mill burned down. Nothing remains of the mill and only a gaping hole marks the mine. (William Metscher collection)

and then built a ten-stamp mill and a 75-ton cyanide plant, which were both completed in 1917. During its first full year in business the company earned a profit of $120,000. Ore from the mine assayed at $40 to $120 a ton.

The company was reorganized in early 1918, when John Kirchen replaced A.G. Raycroft as president. The Morning Glory Mining Company initiated protracted litigation concerning a few rich claims. The dispute was finally resolved in favor of the White Caps Company in May 1918. By 1918 the company had made $100,000 in ore sales but had expenditures of more than $250,000. It dug the main shaft to a depth of 650 feet, but the ore values sank to $10 a ton. The total production value for 1918 was only $87,000, and this downward trend continued as yearly production diminished.

As the shaft went deeper, the ore also became increasingly more difficult to process. Large quantities of arsenic began to appear in the ore. This meant that an oxidizing roast was needed before the ore could be put into the cyanide tanks. Plans were made to change the mill to accommodate the arsenic ore, but the mill closed in January 1920, and the ore was sent to other mills in Manhattan. Miners worked the mine periodically during the next ten years, eventually digging the shaft to a depth of 1,300 feet. But the ore values weren't high enough to offset the milling and shipping costs, and the company closed down.

Another company, the White Caps Extension Mines Company, worked the White Caps area before selling out to the White Caps Mining Company. John Kirchen was the former company's president. The company was incorporated in 1917 and owned thirteen claims adjacent to the White Caps Mining Company's property. The company's most important property was a 500-foot shaft. The property seemed very promising, but the gold ore veins were short, and the company sold out at a loss. The White Caps Mining Company continued operations on the property for a while and then ceased activity when no new major ore bodies were discovered. In July 1925 the company was reorganized and renamed the White Caps Gold Mining Company. It continued working, trying to find new veins. It enjoyed great success on paper and was the largest shipper of ore in the county in 1933. But despite the fact that it produced $208,000 worth of ore that year, its expenses amounted to $202,000. The story was the same in 1934, when the company earned $219,000 and spent $211,000. The company built a flotation mill at the White Caps mine in 1935, but it was destroyed by fire in 1936. The mine remained active until 1940, nevertheless. After the mill burned, the value of production dropped to $53,000 in 1936 and to $70,000 in 1937. Another fire in May 1939 destroyed the hoisting works and was a primary factor in the company's decision to curtail operations in 1940. From 1918 to 1940 the White Caps mine produced just over $2.7 million worth of ore. In May 1952 Mark Yound and A.C. Corlee bought the White Caps Company property. They reopened the mine and revamped the old Manhattan Consolidated mill to treat antimony ore. The operation only lasted for a little while, and not much was produced.

Argus Resources, which reactivated the mine, purchased the White Caps property in the spring of 1980. The Vanderbilt Minerals Company later worked the mine and a small mill. A great deal of exploration and additional work took place, but the property was not producing in the 1990s.

Remains at White Caps are extensive and interesting. Two buildings are in fairly good condition. One is the old assay office, which is still filled with core samples taken from the mine. The other building served as housing for the mine motors. A huge iron wedge-roasting furnace that looks like a lonely sentinel marks the site. At the top of the hill behind the mine are the remains of a house that belonged to one of the company's officers. It was fairly elegant at one time, but the roof collapsed in the late 1970s, and the wood frame is quickly deteriorating. When visiting White Caps, be extremely cautious. The 1,300-foot shaft of the White Caps mine is in the open ground in front of the two remaining buildings and cannot be seen until one is on top of it. The author almost drove into the shaft, avoiding a long, long trip down only by slamming on the brakes. All warnings aside, however, White Caps is a must for ghost town lovers.

Central Nye County

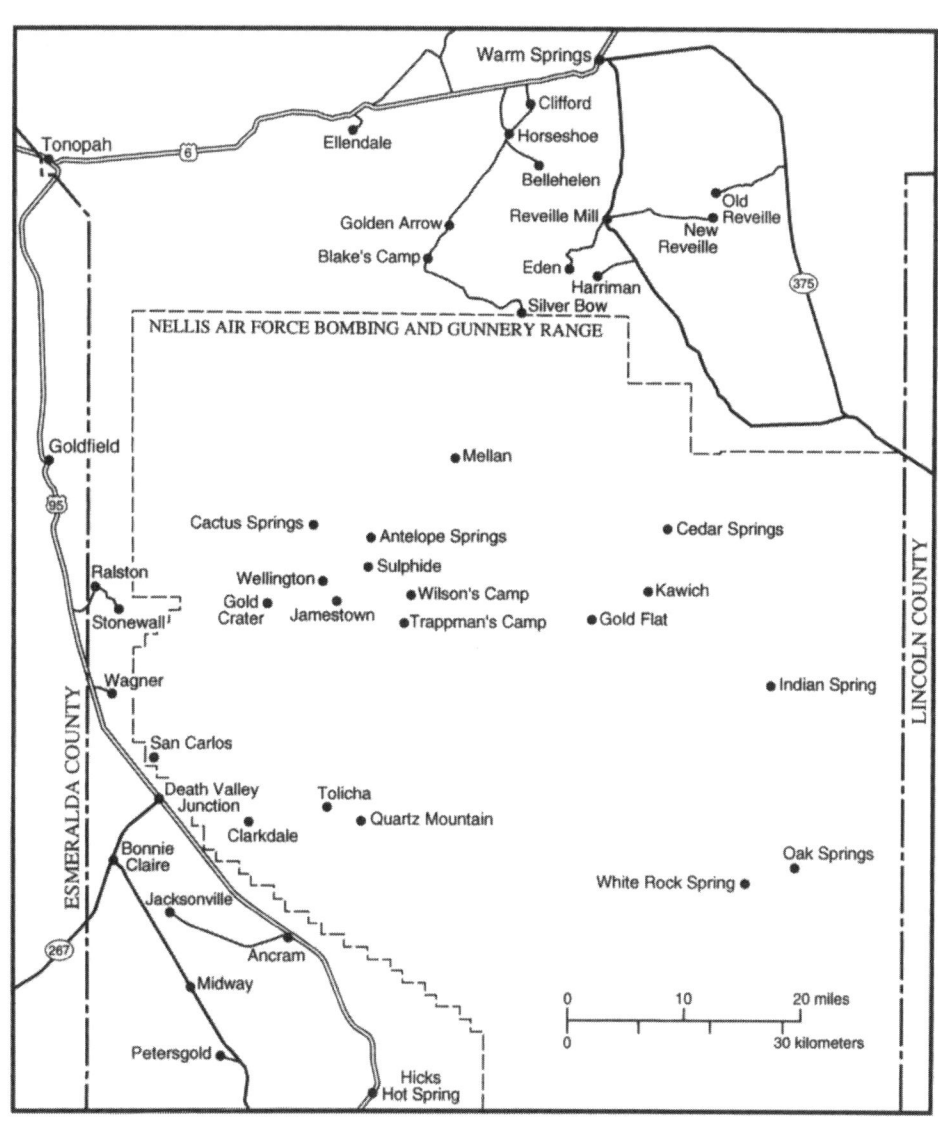

Warm Springs
Clifford
Ellendale
Tonopah
6
Horseshoe
Bellehelen
Golden Arrow
Reveille Mill
Old Reveille
New Reveille
Blake's Camp
Eden
Harriman
Silver Bow
375

NELLIS AIR FORCE BOMBING AND GUNNERY RANGE

Goldfield
95
Mellan

Cactus Springs
Cedar Springs
Antelope Springs
Ralston
Sulphide
Wellington
Wilson's Camp
Kawich
Gold Crater
Jamestown
Stonewall
Trappman's Camp
Gold Flat

Wagner
Indian Spring

San Carlos

Death Valley Junction
Tolicha
Clarkdale
Quartz Mountain

Bonnie Claire
Oak Springs
Jacksonville
White Rock Spring

ESMERALDA COUNTY
267
Ancram
Midway

Petersgold
Hicks Hot Spring

LINCOLN COUNTY

0   10   20 miles
0   30 kilometers

Ancram

DIRECTIONS: *From Beatty, take US 95 north for 16 miles. Exit left onto a faint road, which is the old Bullfrog-Goldfield Railroad right of way. Follow this for 2 miles to Ancram.*

Ancram, a water stop on the Bullfrog-Goldfield Railroad, was active from November 1906 to January 1928. In 1910 and for a few years afterwards a road used for stage travel passed through Ancram, and for a while the site served as a passenger stop. In May 1923, due to the volume of ore being produced at Tolicha, a special siding was created to load the ore. Only a small depot was ever built at Ancram, and today the site is indistinguishable from the surrounding desert.

Antelope Springs

(Monte Cristo)

DIRECTIONS: *Antelope Springs is located 30 miles east-southeast of Goldfield. The site is off limits, as it is inside the Nellis Air Force Bombing and Gunnery Range.*

The Bailey brothers, who had earlier made substantial discoveries at Cactus Springs, initially discovered ore at Antelope Springs in 1903. The brothers staked their claim a mile southwest of Antelope Springs. Antelope Springs was noted for its main spring, which produced 500 gallons daily to provide the district with a commodity that was quite rare in the desert. Pat and Michael Jordan and Frank Reilly made additional discoveries in 1906. The Bailey and Jordan brothers and Reilly were Antelope Springs' only inhabitants until November 1911, when the first semblance of a camp formed after two newcomers discovered rich ore deposits just south of the springs.

The new discovery was named the Antelope View claim. George Wingfield and the Goldfield Consolidated Mines Company paid $15,000 to lease the claim for a week. The company found rich deposits, but when the week was up the owners demanded so much for the property that Wingfield and his company left. Even with Wingfield gone, people still continued to come to the area. In December the Jordan brothers hit ore assaying at $7,000 a ton, and the rush was back on. Within a week 50 people had shown up, and by the end of the year Antelope Springs's population stood at 150. T. M. Johnson platted a townsite. Many buildings were moved from Jamestown, and businesses at Antelope Springs included a lodging house, owned by Casey McDaniel; two saloons; and a restaurant. By the spring of 1912 the camp had a dozen wooden buildings on two separate townsites. The second town, Monte Cristo, located a couple of miles to the west, had formed in 1907 after the Bailey brothers found another ore vein, but the Antelope boom emptied the camp. The boom at Antelope Springs also killed the nearby camp of Trappmans. A stage line to Goldfield was set up, and it operated every other day. The owners were Fred Gleason and John Keefe. In April residents raised $850, $100 of which came from Wingfield, and built a road to Goldfield, which ran through the old camp of Jamestown.

In addition to the Antelope View claims, a new claim group, the Western Union group, was discovered. The Antelope Mines Company, incorporated in March 1912 as a subsidiary of the National Merger Gold Mines Company, also became active in the district. The company controlled three claims, all located just north of the camp. One of the claims, the Antelope mine, had a 300-foot shaft and 1,500 feet of lateral workings. The company never employed more than ten men, but it still managed to produce a respectable amount of good ore.

There were eight other claim groups in the district, and of these the Antelope View group was the most productive. The Antelope View mine had ore that assayed at an average of $200 per ton, with some of it reaching as high as $600 per ton. The mine had an 85-foot shaft with a 150-foot crosscut tunnel. The Chloride group, located 600 feet southeast of the Antelope View mine, had a number of exploratory shafts and tunnels ranging from three feet to fifteen feet, but nothing very valuable was ever found. The Bailey brothers continued to work the oldest group of claims until 1912, but they never found any substantial ore deposits. The other four claim groups (the Auriferous, the Good Luck, the Reflections, and the Star of Hope) were worked on a very low level, and ore values ranged from $6 to $75 per ton.

Difficulty transporting the mined ore to Goldfield hurt Antelope Springs's chances for survival. Even though the ore was extremely rich in some spots, the cost of getting it to mills in Goldfield cut the actual profit margin drastically. After the population peaked in 1912, Antelope Springs headed downhill rapidly. Only the richest ore could be sent to Goldfield, and that was quickly running out. By 1913 the population was under fifty, and by 1914 the district had been completely abandoned.

Activity didn't return to Antelope Springs until November 1925, when Gilbert International bought the mines of the idle Antelope Mines Corporation and began operations. Ore was shipped through the Ralston station on the Tonopah and Goldfield Railroad. In March 1928 the National Merger Company, of which Frank Maloney was president, reentered the district and installed new equipment in the Antelope mine. The company hauled in a large iron building from Jamestown for use as a hoist house, but by 1930 activity had stopped once again. In 1931 Pat McAuliffe, the former superintendent of the Antelope Mines Company, and Al Donohue began developing the Gold Dome and Sulphide claims and found some promising gold ore. In 1933 McAuliffe purchased the Antelope mine, which had produced $73,000 worth of ore, from William Royle. McAuliffe died in July, and activity around Antelope effectively ended upon his death. No revival was ever attempted, and the townsite slowly fell into ruin.

Before being incorporated into the Nellis Air Force Bombing and Gunnery Range in 1950, Antelope Springs was visited by an old-timer from Tonopah. He found two buildings still standing, with the ruins of a number of others scattered around the site. It is not possible to provide an up-to-date report on Antelope Springs, since the site is now off-limits to the public.

Bellehelen

(Henry)

DIRECTIONS: From Tonopah, head east on US 6 for 34 miles. Exit right and follow this road, bearing left, for 10 miles to faint remains of Bellehelen.

Tom O'Donnell, William Whitesides, Al Carter, Andrews, and Nesbitt discovered gold and silver ore on the west flank of the Kawich range in early 1904. The men opened nineteen claims. In September 1906 a townsite called Henry was platted, and a voting precinct was established. Businesses included an eating house, a store, and a saloon. By 1907 the Bellehelen Mining Company had formed and was working the Henry group. The Cornforth property had the richest ore and employed a dozen men. Soon a small camp had formed. A post office, with Captain O. Henry as postmaster, opened at Bellehelen on April 27, 1907, but closed on March 19, 1908. In 1907 George Wingfield and his engineers visited the camp to look at the mining properties. They did not find any lasting ore bodies and soon left. A short-lived newspaper, the *Bellehelen Record*, began publication in April 1907 to promote Bellehelen, but only lasted until the short boom faded in the fall. Very little is known about the newspaper, and no known copies exist. By March 1908 P.V. Meyer was running a store, and there were six active mining companies in Bellehelen. These were the Tomahawk, the Crescent, the Cornforth, the Columbia Nevada, the Sciota, and the Horseshoe. In July the Bellehelen Mining District formed. Sam Sawyer was the recorder. The recently completed Horseshoe mill was enlarged to twenty stamps, but the camp folded by the fall of 1908.

In November the Nevada Bellehelen Mining Company purchased six claims, including the Cornworth mine, near the old camp. Soon almost 500 people were back in the Bellehelen district, feverishly searching all the side canyons for gold. The company had one 600-foot tunnel mine and dug two shafts, one of which was 175 feet and the other 140 feet. The ore per ton ran from $25 to as high as $165, with values running two-thirds silver and one-

third gold. The company rented a ten-stamp mill in Hawes Canyon, located in the Golden Arrow Mining District. The ore from the Bellehelen mines was shipped to this mill for processing. The tunnel was the big producer of the six claims, but the Never-Sweat and the Sciota mines also produced consistently. Between 1909 and 1911 the Nevada Bellehelen Mining Company removed more than $500,000 worth of gold and silver from the Bellehelen district.

The post office reopened on October 15, 1909, and Flora Meyer served as postmaster. Activity began to decline in late 1910, and by the beginning of 1911 the town had only fifty residents. The post office closed on November 15, 1911, and the district fell silent soon afterwards. In 1913 the Sciota-Nevada Mining Company and the Columbus-Nevada Mining Company both restarted their dormant mines. By the end of the year the Nevada-Bellehelen and Tomahawk companies had joined them, but mainly development took place.

The district ended its slumber in 1917 when the Pacific States Mining Company purchased a number of claims near Bellehelen. The company produced more than $100,000 worth of ore before merging with the Tonopah-Kawich Mining Company. In June 1919 T.T. Cornworth and J.B. Wainwright, who had come to Bellehelen during its early days, took over the Tonopah Company. In 1918 the company had produced ore worth $160,000. The merging companies formed a new company, the Bellehelen Merger Mines Company, which was incorporated in 1920. J.W. Oldham was president. The company owned fourteen claims in the Bellehelen district and employed fifty people. A new vein, the Midvale, was discovered in 1920. It carried values of $19 to

$72 per ton. The company built a fifty-ton cyanide mill in late 1922 and early 1923. The mill ran by the countercurrent decantation process. Operations began in May 1923, but the mill was plagued with numerous problems. It shut down several months later, and the company only used it a few times after that. The company ceased operations in the district in 1924, and while it planned to reopen in 1925, plans fell through when financial problems arose.

Another company, the Elgin-Bellehelen Divide Mining Company, became active in the Bellehelen district in the teens. The company was originally known as the Silverfields Ajax Mines Company, but the name was changed in March 1919 during reincorporation. The company purchased a gold and silver mine in the district and immediately began intensive operations. It worked the mine to a depth of 120 feet before operations stopped. The company leased out the property until 1923 and then left the district to concentrate efforts on its property in Esmeralda County's Divide district.

Three other mining companies worked the district in the 1920s. The Bellehelen Extension Mining Company, incorporated in early 1923, had six claims adjacent to the Bellehelen Merger Mines Company's property. Bellehelen Extension had limited production and never realized a large profit. The company folded in 1925 and sold its holdings to the newly formed Bellehelen Development Corporation. After buying the property the company reopened three mines, two of which were vertical shafts and which spanned 120 acres. The best producer was the tunnel mine on the Starlight claim. Its ore assayed as high as $26 per ton. The company faded in 1926 and by 1927 had folded.

The other company to work the district was the Bellehelen Mines Company, which worked the Blue Eagle, the Uncle Sam, and the Ajax mines. The company was renamed the Bellehelen Consolidated Mines Company in June 1924. L. Christenson, who was also the proprietor of the Tonopah Laundry, owned the company. Two new mines—the Bellehelen Queen and Silver Dollar—started up. The company took over six claims in the district, but the operation was doomed to failure, and the company folded in 1927. Despite all the activity that took place in the 1920s, very few businesses opened at the town because most supplies were available at company stores. The only store that was not a company store was the St. Peter general store. In October 1926 the Clifford Gold Mines Company purchased the Bellehelen Merger Company property, refurbished the mill, and hauled its ore from Clifford for treatment. The mill, enlarged to 100 tons, started up in March 1927, and more than 100 people attended the celebration. Ten homes from the early years still stood at Bellehelen and were used for the mill workers. Shortly after the mill started up, Gus Peterson found rich gold in his mine, and the resultant rush brought 400 people back into the district. But Peterson's gold strike was limited, and almost all the new arrivals had departed by the fall. After the Clifford Company ran into financial trouble in late 1927 the mill only operated intermittently until it closed down permanently in 1929. Bellehelen was once again a ghost town. The last bout of activity took place in 1933 and 1934, when C.A. Anderson, the Craig brothers, and R.H. Murphy built a small, three-ton mill at a cost of $500. During their first three months of operation the men made $1,500 reworking the old mine dumps, but by 1934 the work was done, and they left. Bellehelen had emptied for good.

The ruins of Bellehelen are extremely disappointing. Absolutely nothing remains of the camp. Only the ruins of the mill mark the site, and even these are not very interesting. The ruins of the Bellehelen Ranch are half a mile above the mill ruins. The canyon is filled with wild horses, but the site itself is not worth the trouble it takes to find it. An interesting aside about Bellehelen: in the 1920s B.M. Bower wrote a novel about the town called *The Bellehelen Mine*. A few copies of the rare book are in Nevada libraries.

Blake's Camp

(Taylor)

DIRECTIONS: *From Golden Arrow, head southwest for 4 miles to Blake's Camp.*

Blake's Camp was an extremely short-lived mining camp that formed in June 1905 after gold was discovered in a dry wash west of Golden Arrow. The excitement caused by discoveries that prospectors made at Golden Arrow and Silver Bow brought a flock of prospectors to the area to comb every canyon of the district. Blake's Camp soon had a population of about twenty-five. The gold ore was free gold and was in a 600-foot-long, 2-foot-wide area. The ore assayed at only $15 per ton, and soon the camp was nothing but a memory.

Few buildings were erected during Blake's Camp's sixteen-month existence. Some prospectors constructed small shelters, but they weren't anything substantial. It wasn't until May 1922 that new discoveries brought activity back to the district. The Downey brothers, J.D. Maryhouse, and Louis Dulcie discovered the Storm Cloud claim group and quickly bonded it to Frank Taylor and W.A. Sarger. Two shafts were sunk, and assays ran as high as $559 a ton. O.K. Reed, a longtime Nye County miner and rancher whose home ranch was only a few miles away, leased some of the claims and began extensive development. Prospectors quickly claimed all the land within a two-mile radius. The Reed mining district formed, and the first house, which had been moved from Golden Arrow, was erected. George Wingfield visited the district and optioned the Storm Cloud mine but wouldn't pay the owners' high asking price. By June the camp of Taylor contained two frame buildings and three tents, and there were about twenty-five other prospectors scattered through the district. But by July only the Storm Cloud was being worked, and all interest in the district had faded. When work at the Storm Cloud ended in the fall the district was silent once again. No other activity has taken place there since, and only some wood scraps and the collapsed shaft of the Storm Cloud mark the site.

Bonnie Clare
(Clare) (Clair) (Thorp's Wells) (Thorp)
(Montana Station) (Summerville) (Gold Mountain)

DIRECTIONS: *From Beatty, head north on US 95 for 35.4 miles to Scotty's Junction. At the junction, head west on Nevada 267 for 6½ miles. The remains of Bonnie Clare are on the right at the base of the mountain.*

Mining activity began in the Bonnie Clare district in the 1880s, and a small stamp mill was built at a site known then as Thorp's Wells. The mill handled ore from three major mines—the Rattlesnake, the Hard Luck, and the Courbat—all located near Gold Mountain, six miles to the northwest. The ore was transported to the mill by teams of fifteen to twenty mules. The mill operated into the twentieth century and the Bonnie Clare Bullfrog Mining Company purchased it soon after the turn of the century.

A small camp began to form, and soon a stage line from Bullfrog to Goldfield ran through the camp, which is now known as Thorp. A small suburb called Summerville developed one mile to the northeast, but it quickly faded as Thorp continued to grow. In 1904 another mill, the Bonnie Clare, was built to treat ore from all over the district. A small post office, named Thorp, opened on June 15, 1905. Charles Young was postmaster. The camp continued to function on a relatively low level until late September 1906, at which point it received a big boost when the Bullfrog-Goldfield Railroad reached Thorp. The Bullfrog-Goldfield Railroad station was known as Montana Station. However, the residents did not like that name, and when a new townsite was platted in October 1906 the town was renamed Bonnie Clare, after an early settler's daughter. During the next couple of years many attempts were made to establish the Bonnie Clare post office at the new townsite. However, the old Thorp post office remained the only office in the area. The government rescinded a Bonnie Clare office on October 5, 1907, again on March 3, 1908, and once again on June 29, 1908. Finally, on July 13, 1909, the Thorp post office was renamed Bonnie Clare.

A number of new properties on Gold Mountain, which the Nevada Goldfield Mining Company owned, began to produce consistently in November 1906. The original discovery, the Rattlesnake, was the property of the Bonnie Clare Bullfrog Mining Company. The company built a twenty-stamp mill in

November 1906 and employed fifty men. The Bullfrog-Goldfield Railroad soon began to work on a spur line to the mines. A second railroad, the Las Vegas and Tonopah, arrived in August 1907. An impressive two-story wooden depot station was built to serve travelers. Soon after the arrival of the Las Vegas and Tonopah Railroad, Bonnie Clare reached its peak. The town had a population of a little more than 100. Business establishments included saloons, stores, and boardinghouses. Bonnie Clare is one of the few towns that can boast of a bona fide train wreck. On October 1, 1907, an open switch caused the derailment of a Bullfrog-Goldfield Railroad train. No one was killed, but a number of people were injured.

In 1908 and 1909 the combination of Rhyolite's collapse (which slowed railroad traffic) and the decline of the Gold Mountain mines put a damper on Bonnie Clare's hopes for the future. The New Bonnie Clare Mining and Milling Company formed in February 1910 and renovated the old Bonnie Clare mill at a cost of $20,000. In April 1910 the mill was enlarged, and the company built its own spur to its mines on Gold Mountain. The company employed seventy-five men, but the venture proved fruitless and was soon abandoned. By 1911 Bonnie Clare's population had dwindled to less than fifty, and the tailspin continued. Some activity at Bonnie Clare took place in 1913 when the Jumbo Extraction Company built a new mill on the old Thorp property. The mill processed ore from the Happy Kelly mine, located just south of the Bonnie Clare townsite. In 1915 the Quigley Reduction Company took over operation of the mill.

Ruins of a boardinghouse at Bonnie Clare. An old powder house is just to the right. (Shawn Hall collection, Nevada State Museum)

When the Las Vegas and Tonopah Railroad and the Bullfrog-Goldfield Railroad consolidated in 1914 the Bullfrog-Goldfield right of way was abandoned, while the Las Vegas and Tonopah maintained limited operations through Bonnie Clare. The last real railroad service came in June 1927, when Bonnie Clare was a shipping point for all supplies used to build Scotty's Castle. After the railroad folded in 1928 the life quickly ebbed out of Bonnie Clare. Within a short time only the postmaster and a few others were left. When the post office closed on December 31, 1931, the population stood at two. The office had been kept open mainly to handle mail for Death Valley Scotty. Soon after the office closed, only empty buildings faced the hot days and freezing nights, and ghosts reigned supreme over the crumbling ruins.

In 1940 and 1941 the Bonnie Clare Syndicate, of which R.W. Sanderson was president, treated the old dumps and made a fair profit. In December 1951 George Lippincott began construction of a mill and smelter at Bonnie Clare to treat lead ore from his mines in Death Valley. He ran the Sun Battery Company and the Nic-Silver Battery Company of California, and he used the lead for the manufacture of storage batteries. Lippincott spent $250,000 to build the facility and worker housing, and he started the mill in February 1952. After a few years of operation, the mill shut down when Lippincott's mines played out. Bonnie Clare has been completely abandoned ever since, although miners occasionally work the Gold Mountain mines to this day. The ruins of Bonnie Clare are quite interesting and are well worth the trip. A few complete buildings, along with a number of stone ruins, remain. Only wooden rubble marks the site of the once-imposing depot. The ruins of the Bonnie Clare and Lippincott mills dominate the site and are adjacent to the stone ruins. When in the Bonnie Clare area, be sure to take a side trip to Scotty's Castle, located just over the state line in California.

Cactus Springs
(Camp Rockefeller)

DIRECTIONS: Cactus Springs is located 10 miles north of Gold Crater, inside the Nellis Air Force Bombing and Gunnery Range. It is off-limits to the public.

The Cactus Springs district was first worked in 1901 when William Petry came across turquoise deposits on Cactus Peak. Prospectors discovered the largest mine in the district, the Cactus Nevada Silver mine, in 1904. Some placer claims were discovered in 1906 and were mined with limited success for a couple of years. In February 1907 the Plymouth Goldfield Mining Company took over the Davenport claim and started the first serious mining at Cactus Springs, but its success too was limited. The Cactus Range Mining Company was also active in 1909, and in 1910 the Lincoln Gold Mining Company entered the district and built a small camp named Camp Rockefeller at its mine. The Rocket Mining Company, of which John McEachin was president, also began operations in September 1910. After five years of work the Lincoln Company was rewarded with the first rich strike in the district, but

Cactus Springs in 1950, shortly before its incorporation into the bombing range. (William Metscher collection)

the ore was a short vein that disappeared quickly. Mining activity was limited until 1919, when two mining companies became active in the district. The Cactus Nevada Silver Mines Company was incorporated in 1919 under the control of Joseph Nenzel of Rochester, Nevada, who also owned the Nenzel Divide Mining Company. The company purchased the seven Bailey claims, which were discovered when Petry made his discoveries. Bailey had sunk a few shallow shafts during his years of ownership and had exposed a body of silver ore that was over 2,000 feet long and as much as 6 feet wide. The ore Bailey removed assayed at $30 a ton. It was primarily silver but had traces of gold. The new company extended the richest shaft from 150 feet to 500 feet and spent $280,000 to develop the mines and bring in new equipment. The company worked three shifts and built a boardinghouse, a bunkhouse, a blacksmith shop, and an office building. The mine never returned the investment, and the company folded in 1924.

The second company, the Cactus Consolidated Silver Mines Company, was incorporated in 1919, with J. Kendall of Goldfield as president. The company owned the Cactus View claims, which were located to the northeast of Cactus Springs, and while shafts weren't sunk, ore from surface deposits assayed as high as $900 a ton. The company didn't make its first shipment of ore until October 1920. The ore returned $100 a ton, but it wasn't enough to keep the company going. The Cactus Leona Silver Corporation was next to try and make a go of it in Cactus Springs. Cactus Leona bought seven claims in the district in 1920 and spent $7,000 to retimber an old 200-foot incline shaft and drive a 500-foot tunnel. The two mines followed a 10-foot outcrop of ore that returned almost sixty ounces of silver and $10 in gold per ton. The company gave up in 1924 after the ore ran out, and it conducted only scant exploratory work over the next few years.

The last company to work the Cactus Springs area was the Gresham Gold Mining Company, which was incorporated in February 1924. The company bought seven claims on Cactus Peak. The Kennedy Tellurium Mines Company and the Cactus Range Gold Mining Company had formerly owned these claims. Initial ore shipments assayed at more than $50 a ton, and the company sank a 200-foot shaft along with 150 feet of branching drifts. The mine had a sixty-horsepower Fairbanks-Morse hoist and a compressor. The ore body was fairly small, unfortunately, and the company stopped operations in 1927.

Gresham conducted the last organized mining activity in the Cactus Springs district. Prospectors owned two smaller claim groups that they worked for a number of years after Gresham left the district. The larger of the two was the Silver Sulphide group, owned by Edith Bailey of San Bernardino, California. Work had begun on these claims in 1920, and one claim, the Cactus Silver mine, turned out to be a profitable investment. The mine, which was eventually sunk to a depth of 165 feet and which had 800 feet of lateral work, was dewatered in October 1925 after it had been abandoned for twenty years, and a hoisting and pumping plant was installed. The second claim group was the Thompson group, owned by Leonard Thompson. He sank a number of shallow shafts, one of which was 230 feet. These yielded both gold and silver as well as some beautiful turquoise. In May 1932 Paul Savic bought nineteen claims, including the Thompson mine, and began development. T. C. Campbell found gold ore assaying at $149 a ton in his Lucky Truck mine, but all activity ended by 1935, and after producing $100,000 the district was abandoned for good.

Cactus Springs was ideally located. The springs yielded 500 gallons a day, and there was excellent grazing in the area. But that did not save the settlement when the ore ran out. Although properties in the district started out very rich, the holdings almost always quickly panned out without producing substantial amounts of valuable ore. The small camp never had more than fifty people because there never was a great deal of interest in the district. Other than a few boardinghouses and a number of small wood shacks, no other substantial buildings were ever erected there. The site was later incorporated into the Nellis Air Force Bombing and Gunnery Range, isolating the ghosts of Cactus Springs from all human contact.

Cedar Spring

DIRECTIONS: *Cedar Spring is located 12 miles north of Kawich inside the Nellis Air Force Bombing and Gunnery Range. It is off-limits to the public.*

Cedar Spring was a small silver camp that formed soon after the turn of the century. It was located near two good springs, the Jarboe and the Sumner. The camp was still listed on maps as late as 1910, but by then all activity had ceased. No production in the area was ever recorded, and Cedar Spring's population never exceeded fifteen. Because of its location, a present-day status report is not available.

Clarkdale

DIRECTIONS: *Clarkdale is located 1½ miles east of Trappman's Camp, inside the Nellis Air Force Bombing and Gunnery Range. It is off-limits to the public.*

Clarkdale formed after Tom Clark, a big investor at Longstreet, discovered gold in January 1932. After months of quiet development, the announcement of his find led to a rush to the district in August. L. L. Patrick platted a townsite, and homes were moved from Beatty and Tonopah. Both George Wingfield and Tasker Oddie visited the growing camp. In September Clark organized the Clark Gold Mining Company, and the town, which contained five frame buildings and fifteen tents, was added to the Machado Tonopah–Las Vegas stage line. Operations continued to expand during the month of October, and the Clarkdale Extension Mining Company, the Clarkdale Consolidated Gold Mines Company, and the Clarkdale Gold Mines South Extension Company formed. The first ore shipment was made, and a large boardinghouse was completed. However, the excitement quickly died down, and by the summer of 1935 little was being done. In November Ernest Holloway leased the property of the Clarkdale Gold Mines Company and built a twenty-ton mill. He worked the mines but had little success and gave up in 1937. After Holloway left, Clarkdale faded into history and was later incorporated into the Nellis Air Force Bombing and Gunnery Range.

Clifford

(Helena)

DIRECTIONS: *From Warm Springs, head west on US 6 for 7 miles. Exit left and follow the road for 1½ miles to Clifford.*

Contrary to popular belief, the Clifford district was actually discovered in 1905 by an Indian from Tonopah named Johnny Peavine. James and Edward Clifford then staked a few claims, one of which turned out to be the rich Clifford mine. James Clifford did most of the actual work on the prime oxidized ore deposits that were fairly accessible at a depth of less than five feet. The Cliffords sold their claims for $250,000 in 1905. Announcement of the sale stirred up considerable interest in the district, and Clifford began to boom.

Less than two weeks after the sale, a small tent city, which included a few saloons, had sprung up. New strikes around the Clifford area soon brought even more people to the district. Clifford was close to the Ely stage line, and ore mined in the district was shipped forty-two miles west to Tonopah. In 1906 Clifford received a big boost when some very important investors began buying large holdings in the district. Among these investors were Charles Schwab of Pittsburgh and "Diamondfield" Jack Davis of Goldfield. Their involvement seemed to assure the town's future, and W. A. Coyne laid out a townsite in November, selling lots for $100. In December Phillip Meyers completed his large hotel, three more saloons opened, and the camp had a population of 100. In the early days of 1906 new mining companies formed. These companies included the Clifford Hill Mining Company, of which D. F. McCarthy was president, and the Clifford Extension Gold Mining Company,

CLIFFORD MINE, ONE OF THE OLDEST PRODUCERS (

The rich Clifford Mine in the 1910s. (William Metscher collection, Central Nevada Historical Society)

of which J.W. Skelton was president. The Tonopah-Clifford-Reveille stage began running in March. The fare from Tonopah was $6.

The year 1908 marked Clifford's peak. A post office, called Helena, opened on December 8. Wayne Smith was postmaster. The town's population reached 500, and there were 100 houses and numerous tents. The more substantial buildings in Clifford included saloons, a large dance hall, stores, and a number of boardinghouses. The Nevada Broken Hills Mining Company, which had purchased the Clifford mine, built a new freight road to Goldfield. The company's mine produced an average of $10,000 a month worth of ore from November 1908 to June 1909 and was the only producer in the district. In December 1908 J.W. Goodwin started a tri-weekly stage to Tonopah, and by January 1909 Clifford contained 125 buildings, and the Helena Townsite Company had completed a telephone line to Tonopah.

The Clifford mine slowly ran out of good ore, and the town was soon dying. The post office closed on July 15, 1909, signaling the end of Clifford. The ghosts of Clifford were disturbed only by leasers who continued to work the area, although the Clifford school remained open until 1912. Jim Clifford, who had regained control of his mine by default, continued exploring and in February 1913 found ore that returned $863 a ton. He continued to make occasional shipments and built a small mill at the mine in the spring of 1914. In May 1919 Joseph Shea and Briz Putnam bought the mine and later, in January 1921, sold it to Howard Broughton for $100,000. He organized the Clifford Silver Mines, Inc., but it did little work.

In April 1925 the Gilbert Clifford Gold Mines Company, which changed its name to the Clifford Gold Mines Company in July 1926, purchased the Clifford mine and fourteen claims around the mine for $150,000, and new operations began. This brought new life to the empty town of Clifford. Many houses were moved to the town, and the school district was reestablished. In

June 1926 the old consolidated Spanish Belt mill at Barcelona was dismantled and used to enlarge the Bellehelen mill, which the company had purchased. The mine had been under option to the Clifford Silver Mines Company. The new company sank four shafts in the Clifford area. All the shafts were about 300 feet deep and followed a new lode about 50 feet wide. The ore initially removed was worth an average of $50 a ton. A diesel generator supplied electric power, and a Chicago pneumatic compressor was also pressed into service. The ore had to be shipped to Tonopah until the Bellehelen mill was rehabilitated and restarted in 1927. Then the company sent ore to that mill for processing. Some shipments assayed as high as $550 a ton. However, the company was attached for debts in September 1927. Operations were limited after that, and the company folded in 1929. Clifford emptied one last time, and the buildings were eventually moved elsewhere. Except for a small-scale revival near the end of the Depression, the district has been quiet ever since. It produced a total of $500,000 worth of ore, and some estimates of its total production value run as high as $750,000.

Not much remains of once-bustling Clifford. A fire in August 1946 destroyed the last two buildings at the townsite. The site is marked by the huge tailing pile of the Clifford mine, which is clearly visible from US 6. Only one building of unknown origin remains at the site. Some of the mine's workings are left, but otherwise nothing substantial is left of the town except for piles of rubble. Clifford is not one of the better ghost towns, but it did play an important part in Nye County's history.

Death Valley Junction

(Scotty's Junction)

DIRECTIONS: *Death Valley Junction is located 6½ miles east of Bonnie Clare at the junction of US 95 and Nevada 374.*

For many years Scotty's Junction was nothing more than the turn-off for the road heading to Death Valley and Scotty's Castle. In the late 1920s the Pacific Coast Borax Company built a mill at the junction and constructed a company town called Death Valley Junction. The town did not contain any saloons or dance halls, per the company's orders. Because of the location, the town and mill virtually closed down during the hot summer months. At its peak, 100 people lived in the town. The Great Depression forced the company to fold in the 1930s, and the town was dismantled. For many years, a gas station, a restaurant, and small motel operated at the junction until it was destroyed by fire in the 1970s.

Eden
(Gold Belt)

DIRECTIONS: From Warm Springs, head east on Nevada 375 for 1½ miles. Exit right and follow this road for 13 miles. Exit right and follow the road for 2 miles to the Eden Creek Ranch. Continue past the ranch for 4 miles to Eden.

The Eden district, which was also called Gold Belt, formed after John Adams and Joe Hardy discovered silver and gold ore in January 1905 at the head of Eden Creek, then known as Little Mill Creek. By the end of summer the camp had a population of forty. In July 1906 the Nevada Gold Sight Mining Company, of which M.T. Rowland was president, purchased a few claims and began development of the Silverfield mine. But by the summer of 1907 interest faded, and soon Adams was the only one left. Adams worked his find for a number of years but finally sold the property to Mark Bradshaw, who organized the Eden Creek Mining and Milling Company. In December 1921 the company moved six houses to the operation. The company continued to work the Adams mine, which followed the Eden vein for almost 300 feet. The ore from the mine contained thirty ounces of silver and half an ounce of gold per ton. The company employed ten men, but the small camp actually had a population of around twenty-five, because many others were prospecting the sides of the canyon for new deposits. The company was reorganized in late 1922, and J.W.S. Butler of Sacramento, California, became

The Terrell Mill at Eden Creek in the 1930s. With hand tools Guy Burch, a Tonopah carpenter, built the mill. (Steve Balliet collection, Central Nevada Historical Society)

president. The new management began work on a tunnel. By June 1923 it was 1,400 feet long, with 1,300 feet of drifts.

Another mining company became active in the Eden district in 1925. T. F. Branigan of Tonopah was president of the Crucible Gold Mining and Milling Company, which was incorporated in 1925 with $1.5 million in capital. The company, which owned five claims at Eden, including the Golden Eagle mine, operated on an extremely low level. For example, the company spent a mere $1,200 in 1925 for operating expenses. The company employed only two men, and they worked on two sixty-foot tunnels. The ore from the two tunnels was free milling gold ore but only assayed at $14 per ton. In 1925 the Fallini Silver Mining Company also began operations.

The Southgold Nevada Mines Company, of which K. M. Terrell was president, moved into the district in 1926. Southgold Nevada had a number of claims adjacent to the Crucible Company. The main producer for the company was the Southgold mine, which had a 400-foot tunnel with 1,500 feet of lateral workings. There were also a number of other tunnels around the Southgold mine. The ore from all of these mines assayed at an average of $10 per ton. In 1928 the company built a seventy-five-ton plant at the mouth of Eden Creek. The plant was connected to the mine by a 1,100-foot tramway. The company used this plant to extract $50,000 worth of gold.

Southgold Nevada took over the Crucible Company in February 1930, by which time more than thirty-five miners were working the district. The company leased some of its property to the Oro Cache Mining and Milling Company, which formed in 1929 and which was run by Harriet Stingley. After the company dug a fifty-foot shaft, the likelihood of finding valuable ore seemed small. The company struggled for a few years but finally folded in 1934. Meanwhile, Southgold continued to be active. It considered building a mill in the early 1930s, but it was not until 1938 that a thirty-five-ton mill was actually completed in Eden. Southgold remained active until World War II, and after the company left the district, Eden was emptied forever.

The Golden Crown claim group, which included the old Eden mine, was an important part of the Eden Creek Mining District. The claim group, along with the Southgold and the Crucible, was active during the 1920s and 1930s. The Eden mine had originally been worked in 1906, but hopes of finding new ore deposits in the mine quickly faded, and the group was not worked again after 1935. Another group of claims, the Nevada Triumph (owned by George Chubey), was active in 1926 and encompassed seventeen claims. When Chubey found nothing of value after a year of work, he left Eden. During Eden's many years of mining activity, a large ranch also operated. G. B. Fallini established the Eden Creek Ranch in 1910. He died in 1941 and was buried in the family plot at the ranch. While the ranch is now abandoned, the Fallini family still runs a prominent ranching operation in the area.

Eden's remains are fairly extensive. Four wooden buildings are at the site, as are the scattered ruins of a number of other buildings. The remnants of the tram and wash plant are easy to find, since they are visible from the road. The Eden Creek Ranch, which was occupied until recently, is at the mouth of Eden Creek. The site is well worth the trip, but bring extra gas if you plan to visit, because it is far from any filling station.

Ellendale
(Monitor) (Lucky Strike) (Electrum) (Perkup)

DIRECTIONS: *From Tonopah, head east on US 6 for 26 miles. Exit right and follow this road for 2 miles. Take the left fork and continue for 2 more miles. At the next fork, take a right and continue for ¾ mile. Bear right at the next fork and follow the road for ½ mile to Ellendale.*

In April 1909 Ellen Clifford Nay, for whom Ellendale was named, discovered a large deposit of float gold in the southern part of Saulsbury Wash. The free gold ore was some of the richest ever found in Nevada. At the time, Joe and Ellen Nay were running a stopover place in Saulsbury Wash on what is now US 6. In 1907 they had built a lodging house and a small store, and they provided meals. The district looked very promising, and the rush to the area was comparable, on a smaller scale, to the rushes at Goldfield and Manhattan after prospectors made initial discoveries there. Within a matter of weeks 400 people were in the district. The first shipments of ore reportedly assayed at $80,000 a ton, although official assay reports list the value at $4,100 per ton, which is still an impressive figure. The test load consisted of five tons of ore and returned 205 ounces of gold and 145 ounces of silver per ton. The complete shipment was worth more than $20,000, and the district's production value in 1909 was more than $50,000.

The Ellendale townsite was laid out in June, and almost all the lots sold in the first two months. Regular lots went for $150, and corner ones were $200. More than 100 people wanted the four corner lots, and the winners were determined by a throw of the dice. A.C. Wooley bought the first lot and opened a saloon just hours later. Within one week, twenty buildings were under construction. Many other wooden buildings were built, including a number of saloons and a few boardinghouses for miners. The value and the quantity of gold seemed to suggest that Ellendale was there to stay. By July Ellendale's population had stabilized at 350. Its active mines included the Gold Wedge, the Mustang, the Sandstorm, the Tiger, and the Bernice. A railroad to Tonopah became a possibility, and a group of surveyors even mapped out a route for the railroad bed. A telegraph line—a luxury shared by few other towns at the time—to Tonopah was constructed in late 1909. Two weekly newspapers, the *Lode* and the *Ellendale Star,* began publication. Both were printed in Tonopah and brought to Ellendale by wagon, but neither lasted into 1910.

Ellendale suffered a bitter blow when, in late 1909, newspapers in Rhyolite and Tonopah called its riches a swindle. In fact, doubts had begun to arise as early as June, when people started to suspect that the grand opening of the Ellendale townsite had been staged just to sell lots, since little effort was being made to promote the merits of the local mineral industry. The very ownership of the mine came into question when J.J. Bunch came forward and claimed to have discovered the property, but the Cliffords and the Nays ran him off and had him committed to an asylum in Carson City. He filed suit, but the accusations were proven groundless. The controversy halted the flow of people to the district, but Ellendale still survived. In 1910, when mainly leasers worked the district, its mines produced $45,000 worth of gold and silver. The original claims played out in 1911, but some smaller ore bodies were discovered, and limited production continued through 1912. Total production value for the district from 1909 to 1912 was $100,000.

The Northern Saloon at Ellendale in June 1909. (Nevada Historical Society)

After 1912 ghosts were the district's main inhabitants. The Cliffords, who continued to prospect the area, were the only exception. Some activity took place beginning in November 1929, when the Raymond Van Ness Mining Company discovered a barium ore deposit that contained almost 700,000 tons. At around the same time A. T. Wilkerson and Paul Tucker relocated four claims and discovered copper in the Lucky Strike. A new camp, called Monitor, sprang up in 1926 after M. E. Bailey discovered a rich lead deposit in the summer. About twenty people called Monitor home until the ore body was mined out in 1928, and then the camp vanished. However, work continued on the Lucky Strike, and in March 1930 miners struck a vein of $372-a-ton ore. Soon seven leasers were working the area. As a result, a small new camp called Lucky Strike began to form just below the mine. In April George Wingfield took an option on the Lucky Strike but wasn't willing to pay the price the owners were asking. The camp, located three miles south of old Ellendale, was renamed Electrum in May, when Wilkerson organized the Electrum Gold Mines Corporation. But all work stopped in July when James and Bert Murphy filed suit against Wilkerson, claiming the Lucky Strike was actually theirs. Although the Lucky Strike was silent, other activity was taking place elsewhere in the Ellendale district. The Van Ness Company, which had a large cinnabar operation near Belmont, began shipment from its barium deposit after completing extensive development. Meanwhile, Wilkerson won his court battle in December, but the Murphys filed for a new trial in February 1931. Tom Kendall, Harry Springer, and Lloyd Wilson began working a claim adjacent to the Lucky Strike, and the trio built a small treatment plant in Saulsbury Wash, four miles to the east.

Wilkerson, finally vindicated, resumed work on the Lucky Strike, installed a gallows frame, and hired double crews. At around the same time Jack Clark made a new strike about three miles away, and a new camp, called Perkup, quickly materialized. The flurry of activity was unwarranted, however, as all the mines quickly panned out, and the camps were abandoned by the end of 1931. Only the Van Ness Company remained, making occasional barium shipments. The mine was later taken over by the Glidden Paint Company, which used the ore in its paints, but it was idle by 1937. The Chemical and Pigment Company began making barium shipments again in 1939. The company employed five and was soon shipping 3,000 tons of barite a month. While the deposit was one of the richest in the country, its size was limited, and it was mined out in 1949. The only other activity in Ellendale took place in 1938 and 1939, when the Gilbert brothers reopened the old Ellendale mine and took out $10,000 in gold and silver before the vein they were following panned out. Since then the only inhabitants of the town have been the wind and desert animals, except in the early 1980s when Geo Drilling Fluids, Inc., extracted fair amounts of barite from the Jumbo mine.

Hardly anything remains at Ellendale. For many years, only one building stood at the site, and it finally succumbed in the late 1960s. Apparently the sides of the building gave out and made the roof fall down, because the roof is still completely intact. The sites of the saloons are easy to locate thanks to the broken beer bottles. The only other markers are the depressions in the sand where other buildings once stood. Mine ruins and tailing piles are in the hills surrounding the site. The road to Ellendale tends to be very sandy and should not be attempted without a four-wheel-drive vehicle.

Gold Crater

DIRECTIONS: *Gold Crater is located 10 miles east of Stonewall Mountain, inside the Nellis Air Force Bombing and Gunnery Range. It is off-limits to the public.*

Discoveries on Pahute Mesa in May 1904 led to the establishment of Gold Crater. By September 1904 more than 200 miners had flocked to the booming camp, but most soon left. During the short boom 300 lots were sold in the townsite, most during a two-week period. The majority of these sold for $50 each. This was a low price compared to what lots were going for in other towns, and thus Gold Crater was attractive to speculators, but the town never developed enough for them to capitalize on their investments.

The camp had a population of seventy and contained two saloons, a store, and thirty tents. The small camp elected a mayor, Harry Marow. He was a deputy sheriff who was later murdered in Beatty in 1905.

During the height of the excitement over Gold Crater, a water pipeline was planned from Stonewall Mountain, ten miles to the east, but the boom subsided before the pipeline could be built. Water came from two wells located some distance from town and also from water tanks on Pahute Mesa. The town was almost empty during the winter of 1904. Only a few hardy souls were willing to brave the chilly winter. The following year some leasers came back to work claims in Gold Crater. In March the Gold Crater Mining Company, of which T. D. Murphy was president, formed and worked the district's main mine, the Hub. The company sank new shafts and continued limited development of the mining district. The ore taken out of Gold Crater during this period ran from $40 to $240 a ton and was laced with heavy values of gold and smaller amounts of silver. The leasers left Gold Crater soon after 1905, and the district remained quiet until 1914.

In 1914 the Gold Prince Mining and Leasing Company was incorporated with $500,000 in capital. Dr. A. Farnsworth was president of the company, which was based in Grand Island, Nebraska. It purchased the property of the defunct Gold Crater Construction and Mining Company and built a twenty-ton amalgamation and concentration mill. The gold ore that the Gold Prince Company removed from the mines was low in value, running only $15 per ton. Operations continued, however, and the company installed a twelve-horsepower hoist, but the ore kept on decreasing in value. The mine closed when the shaft was more than 265 feet deep. Despite the shaft's depth, the company removed only a small amount of gold and silver. The operation never reopened, and soon Gold Crater belonged to the ghosts. The only other activity that took place there began in the late 1930s, when Pius Kaelin conducted a one-man operation and also built a two-stamp mill. This bout of activity ended when the site became part of the bombing range. Gold Crater produced a total of $100,000 worth of ore. Now only the rumble of fighter jets disturbs the town's ghosts.

Golden Arrow

(Longstreet)

DIRECTIONS: *From Tonopah, head east on US 6 for 34 miles. Exit left and follow the road for 12 miles to Golden Arrow.*

Discoveries in August 1905 made by Claudet and Marl Page, both of whom were deaf-mutes, prompted a small rush to the Golden Arrow area. The two men sold their claims for $45,000 in January to Eltman and

One of the many active mines at Golden Arrow during 1908. (William Metscher collection)

Scott, and the Golden Arrow Mining Company, of which G.A. Atwood was president, laid out a townsite in 1906. Soon four mining companies, along with many individual claim workers, were in the district. The Golden Arrow Mining District formed in February with J.F. Hall as recorder. In March the Consolidated Golden Arrow Mining Company, of which B.B. Gillies was president, began operations on the Confidence mine. A new two-story hotel was filled to capacity, and the camp's population had reached 200 by the end of the month. A mercantile store owned by C.A. Wood and a restaurant opened, along with at least six saloons. With new arrivals coming every day, the Golden Arrow Townsite Company was established. By 1907 the town had a number of frame buildings. A spring on the Longstreet Ranch, seven miles to the northeast, provided water. A railroad that was being planned from Tonopah to Ely was supposed to run through Golden Arrow, but the idea was dropped when Golden Arrow faded before the plans were put into motion.

The largest of the four mining companies working the district was the Golden Arrow Mining Company. The company owned seven claims, five of which had shafts ranging from 50 feet to 150 feet, and the ore, which carried values in both gold and silver, assayed as high as $100 per ton. Another important company in the district was the Cotter Mines Company, which owned thirteen claims, employed thirty, and controlled the deepest shaft in the district, the Gold Bar, which eventually reached a depth of 500 feet. The ore from the mine assayed at an average of $25 per ton, and a number of ore shipments were sent to the West End mill in Tonopah. The Kawich Consolidated Company controlled twenty-six claims in Golden Arrow and, in terms of acreage, was the largest company in the district. The main producer was the Mascot mine, whose ore assayed from $25 to $55 per ton. The Golden Arrow Mohawk Mining Company, which owned six claims, had a 500-foot tunnel that produced silver and gold ore assaying at an average of $20 per ton. There were eleven other mines in the district, including the Geneva, the Crescent, the Black Dog, and the Bunker Hill. They were not large producers and, for the most part, were independently owned. The Clifford Mining Company also worked the district. The Carlson Mining and Milling Company built a fifty-ton mill in nearby Hawes Canyon in 1907, but a cloudburst in August 1908 washed out the foundation, and it was seldom used afterwards. The Golden Arrow school district was established in December. It had fifteen students, and Della Gilbert was the teacher.

Unfortunately for Golden Arrow, the town was more of a real estate promotion than a profitable mining camp. The ore did not last long enough for a large town to form. Golden Arrow was practically empty by 1909, and the only active companies were the Golden Arrow, the Michigan-Ely, the Cotter Mines, and the Kawich Consolidated. Litigation, which arose as the result of the flurry of mining deals made in 1907 and 1908, played a major role in stalling mines' development. A short revival in 1911 brought forty-five people back to the district, and the Cotter Mines Company and the Golden Arrow Development Company were actively mining. But by September 1912 there were only eight people left in Golden Arrow. After that Golden Arrow was left to the ghosts, with the gaunt remains of the hotel standing guard. In 1913 only the Golden Arrow Development Company, of which C. M. Kieron was president, was carrying on limited activity. By August 1914 the company had spent $152,000 on development and had begun construction of a ten-stamp, 200-ton mill, but financial problems prevented the mill's completion and forced the company into bankruptcy. It wasn't until Kieron returned in June 1920 and reopened some mines that Golden Arrow attracted any more attention. He organized the Golden Arrow Mining Company but was basically the only employee. When he died in the late 1920s the company ceased to exist. In 1931 Cotter returned and reopened a couple of mines. He was the town's only resident. He finally gave up for good in the late 1930s, and Golden Arrow was completely abandoned. During World War II Golden Arrow saw some limited activity, and the Cliffords, from Stone Cabin Ranch, worked the Jeep group in the winters from 1941 to 1946, when ranching was slow. They dug a 100-foot shaft. Before the Cliffords stopped working the group in November 1946 they had mined almost $5,000 worth of gold and silver.

The mines remained quiet until the 1970s. The same company that also re-worked tailings at Reveille and Keystone—the Golden Arrow and Keystone—

began reprocessing the old tailing piles with a cyanide leaching system. Unfortunately, the company was not respectful of history and flattened most of the ruins at Golden Arrow. The company completed its operations in the 1980s and then left the district. There are a number of mine ruins and a few old wood buildings still in the area.

Gold Flat

(Nixon)

DIRECTIONS: *Gold Flat is located three miles west of Quartzite Mountain inside the Nellis Air Force Bombing and Gunnery Range. It is off-limits to the public.*

Gold Flat was one of the many flash-in-the-pan towns located around Pahute Mesa. All these camps formed just after the turn of the century as a result of the excitement at Gold Crater. The small camp was originally known as Nixon, after Senator George Nixon, who had formed the Nixon Townsite Company with partners Tasker Oddie, George Wingfield, and "Diamondfield" Jack Davis. The company also owned thirty-two claims nearby, but the townsite plans flopped when the claims proved to be worthless, and the focus of interest shifted to Kawich, located only four miles away. A few months later, in February 1905, the town's name was changed to Gold Flat in an attempt to show how rich it was. But the name had more gold than did the surrounding hills, and Gold Flat soon ceased to exist. Since the camp never consisted of more than tents, it is doubtful that anything is left there today.

Harriman

DIRECTIONS: *From Eden, travel 2 miles south to Harriman.*

Harriman formed following exploration that occurred as a result of discoveries made at Eden Creek, two miles to the north. There were six major claims at Harriman, all of which the Nevada Gold Sight Mining Company purchased in 1907 and 1908. The company dug three tunnels, which had 1,400 feet of workings, and the gold ore was valued from $10 to $25 per ton.

The camp never grew to any size and at its peak had fewer than twenty-five residents. The mining company built a few wood boardinghouses, but that was the extent of the construction. Supplies were brought in from Eden. The camp did have an ample amount of wood, used both for heating during the cold nights and for support beams in the tunnels. Water, a commodity valued everywhere in dry Nye County, was quite abundant. But the ore was not as plentiful as the wood and water were, and it became increasingly difficult to find. By the end of 1910 the Nevada Gold Sight Company was in the red, and it folded in early 1911. There were still five prospectors digging around Harriman in 1911, but by 1912 the site belonged to the ghosts.

The remains of Harriman are scant. Only the ruins of one boardinghouse mark the site. Fairly large tailing dumps are located on the sides of the canyon, and many small "prospector holes" pockmark most of the area. Harriman is quite difficult to reach, not because of the roads but because of its distance from a gas station. If you plan to visit the site, be sure to bring extra gas.

Hick's Hot Springs

DIRECTIONS: *From Beatty, head north on US 95 for 5.3 miles to Hick's Hot Springs.*

Hick's Hot Springs came into existence in January 1907 when the Bullfrog-Goldfield Railroad set up a grading camp there. The site was later a water stop until the railroad closed in 1928. It is still an active area with a number of hot spring, resort-type setups.

Horseshoe

DIRECTIONS: *From Tonopah, head east on US 6 for 35 miles. Exit right and follow the road for 1½ miles. Take a right at the fork and follow the road straight through for 5 miles. At the end of the road, take a left and follow this road for 1¼ miles to Horseshoe.*

Horseshoe, a small offshoot camp of Clifford, formed in 1908 and had two major mines, the Lawrence and the Original Horseshoe. The Lawrence mine was 200 feet deep and had 750 feet of lateral work. The Original Horseshoe mine was much smaller. After rich initial silver discoveries were made, the ore quickly faded, and the mines were abandoned. The camp died in 1911, and nothing at all remains.

Indian Spring

DIRECTIONS: Indian Spring is located 10 miles southeast of Kawich, inside the Nellis Air Force Bombing and Gunnery Range. It is off-limits to the public.

Indian Spring, a small mining camp, formed at the base of Wheelbarrow Peak. The camp was active during the early 1900s but never achieved any prominence. No more than fifteen residents lived there at any one time, and most of those who did were housed in tents, although a few stone cabins were built. The camp was completely abandoned in 1910 after the two small mines on Wheelbarrow Peak closed. A report on Indian Spring's present condition cannot be made because the site is off-limits.

Jacksonville

DIRECTIONS: From Ancram, continue west on the unimproved road for about 5 miles to Jacksonville.

Jacksonville, a water stop on the Bullfrog-Goldfield Railroad, was active from November 1906 to January 1928. The water stop was in the middle of the hot, dry Sarcobatus Flat. Steam engines were in constant danger of overheating, and the need for water led to the establishment of a number of water stops similar to Jacksonville throughout the flat. The water at the Jacksonville stop came from two nearby wells, the Tonopah and the Seattle. When the Bullfrog-Goldfield Railroad folded in January 1928, Jacksonville's service was no longer needed, and the site soon entered the state of ghosthood. Nothing at all remains.

Jamestown

DIRECTIONS: Jamestown is located 12 miles south of Antelope Springs, inside the Nellis Air Force Bombing and Gunnery Range. It is off-limits to the public.

Jamestown formed in January 1908 after the James brothers, C. S. Hellman, and George McKenneane discovered gold on Pahute Mesa. Soon close to 100 prospectors had flocked to Jamestown. A townsite was platted in April, and within two hours all the main street lots had sold. By the end of the month the booming camp had a population of 400. Water was hauled to the camp from Antelope Springs. In May the *Jamestown News* reported that the town had four stores, five saloons, real estate offices, seven restaurants, two hotels, bakeries, assayers, building contractors, and a stage line

to Goldfield. However, businesses were housed in tents, and only one frame
building had been built. The paper's publisher and editor was J. Mastella
LeGrand. It was first published in May, but there is no further mention of
it anywhere, and it seems as if it must have folded late in the summer, after
its promotional aims had been accomplished. During this boom period Fred
Schultz was one of the town's main developers. He was also the last resident of the district, remaining there until the 1930s. A post office opened on
June 15, 1908, and by the end of summer Jamestown was taking on a more
permanent look, as some of the tents had been replaced by frame buildings.
The Golden Chariot Mining Company, of which Carl Feutsch was president,
was organized in Jamestown in 1908 and began intensive operations. Many
of the claims were copper producers, but there were a few gold mines. The
richest of the gold lodes was the Golden Chariot No. 1, which was more
than 300 feet deep and had initial assays of $200 per ton. Other gold mines
included the Franz Hammel, the Mohawk, the Daisy, and the Last Chance,
all of which Engrace LaBarthe owned.

Jamestown's mines produced consistently only into 1910. The post office
closed on August 31, and the town continued to slip downhill. The Golden
Chariot Company folded in late 1910, and by 1911 Jamestown had emptied.
Some copper ore was discovered in the area in 1916, and in April Feutsch
reactivated the Golden Chariot Company. However, after a couple of years
of only limited success, he gave up once again. A number of buildings were
erected in Jamestown, but not much is known about their origin. A visitor
to the site in the 1930s reported that three buildings, including the old post

office, were standing. Now the site is inside the Nellis Air Force Bombing and Gunnery Range, and only the occasional rumble of Air Force jets disturbs the ghosts of Jamestown.

Kawich
(Gold Reed)

DIRECTIONS: *Kawich is located 70 miles southeast of Tonopah, inside the Nellis Air Force Bombing and Gunnery Range. It is off-limits to the public.*

A small group of prospectors from Tonopah, led by O.K. Reed and Jack May, made initial discoveries of gold ore in the Kawich Mountains during the summer of 1903, but it wasn't until early 1905 that a camp began to form. Water was extremely scarce in the area, and newspapers continually warned readers about how dangerous Kawich was, because it was fifteen miles from the nearest source of water. Instead of scaring people away, the publicity convinced prospectors that the discoveries at Kawich must be incredibly rich, since the papers were trying to discourage them from coming there. Within weeks 400 men had arrived, and a townsite, called Gold Reed, after O.K. Reed, was laid out. Lots ranged in price from $60 to $125, and $10,000 worth of them were sold. By February one hundred tents and one

Failed mill at Kawich. (Central Nevada Historical Society)

Preserving the Glory Days

frame house were huddled together in Gold Reed, which contained eight saloons, three stores, three restaurants, and three lodging tents.

The camp was later renamed Kawich, after Chief Kawich, a Shoshone whose name meant "mountain." Water was brought in from a new source, Cliff Spring, twelve miles to the east. A post office opened on April 10, 1905. Rose Maude Hanley, who also ran a restaurant in town, was postmaster. Despite all the people living in Kawich, the office did only $21 worth of business during the first year. A stage line was set up between Kawich and Tonopah, and a round-trip ticket cost $10. Building continued in the spring of 1905. The Nye and Esmeralda County Mercantile built a large store, one of the most substantial buildings erected at the town. The Kawich Improvement Club formed and had sixty-eight members. The Kawich Gold Mining Company, of which George Nixon was president, and the Kawich Quartzite Mining Company, of which Thomas Fleming was president, both formed.

The Gold Reed Mining Company also became active in the district in early 1905 and began work on three mines: the Chief Kawich, the Gold Reed, and the Diamond. The Chief Kawich and the Gold Reed, both of which were more than 150 feet deep, were larger than the Diamond. The Diamond, however, while it was only 100 feet deep, had the richest ore, although it was limited in amount. The gold occurred in thin pyrite veins that assayed at an average of $35 per ton, but the ore value from the other two mines rarely exceeded $10 a ton. The grade of the district's ore became even lower as the spring of 1905 progressed, and by the end of the summer only ten people were still working the mines. A marriage brightened the spirits of Kawich residents for a while. The postmaster, Maude Hanley, married O. K. Reed in September, and the few people left in town engaged in a big celebration. In November the Nevada Mining and Smelting Company purchased all the mines, but due to the low grade of the ore, activity was limited. Finally, in 1907, the company curtailed its operations to concentrate on its holdings in Tybo.

Kawich was on a downhill slide after only a short existence, and the trend continued until the district was totally abandoned in 1908. The post office operated until June 15, 1908, and the last postmaster, Albert Tait, was one of only three people still living in the town. In March 1929, after it had been abandoned for many years, Kawich revived for a short time when a new strike created a small rush, but the revival fizzled by the end of summer. It was not until the 1940s that anyone came back to the Kawich district. Albert and Robert Martel reopened the Gold Reed mine and dug it to a depth of 300 feet, but upon the formation of the bombing range in 1948 the Martels were forced off the site and were never able to get permission from the government to continue mining. There were no other large-scale revivals in the Kawich district, and a good many people wonder why any activity ever took place there at all. The site is now inside the Nellis Air Force Bombing and Gunnery

Range, isolating Kawich from visitors. It is doubtful that anyone would want to visit the desolate site, anyway.

Mellan

(Queen City)

DIRECTIONS: *Mellan is located 34 miles east of Goldfield, inside the Nellis Air Force Bombing and Gunnery Range. It is off-limits to the public.*

Mellan was a very small mining camp that never produced much and yet managed to last more than ten years. It was founded in March 1930, when Jess and Hazel Mellan discovered gold and quicksilver in the nearby mountains. The ore assayed from $17 to $25 per ton. A group of about fifteen people formed a camp near the discovery site. Mellan was located less than a mile from the old camp of Queen City, which had flared up briefly during the summer of 1905.

The Mellan Gold Mines group gained control of most of the claims and began work on two shafts. During this time, Mellan's population reached a peak of twenty-five. The Mellan Group continued to work the mines until World War II. By then the shafts were 400 feet and 100 feet deep and had almost 700 feet of lateral work apiece. The total production value of the Mellan district was only about $5,000. A small amount of activity took place there after the war, but by 1950 the camp was abandoned. In the 1950s Victor Kral reported that several buildings, some head frames and hoist houses, and an ore bin still remained at the camp. Now Mellan is incorporated into the Nellis Air Force Bombing and Gunnery Range, and it is not possible to report on the camp's present condition.

Midway

DIRECTIONS: *From Bonnie Clare, head back down to Nevada 267. Continue across Nevada 267 and follow this poor road for 1¼ miles. Take the right fork and follow the road for approximately 8 miles to Midway.*

Midway was a small water stop on the Las Vegas and Tonopah Railroad. The stop was used from August 1907 to early 1914. Soon afterward the railroad was abandoned, and the water stop was dismantled. Today nothing remains at the site.

Oak Springs

DIRECTIONS: Oak Springs is located 42 miles east of Tolicha, inside the Nellis Air Force Bombing and Gunnery Range. It is off-limits to the public.

The Oak Springs area was the scene of limited activity during the summer of 1905. A number of claims opened up, and miners removed small amounts of gold, silver, copper, and chrysocolla. The mining activity didn't last long, and the area was not active again until 1911, when the Oak Springs Copper Company began operations. The company, based in Provo, Utah, formed on May 20, 1911. E.K. Ferguson was president. The company purchased the Washoe claims. Oak Springs Copper worked the site extensively for one and a half years and shipped out a fair amount of copper ore. The claims ran dry in early 1913, and the company pulled out of the district.

Oak Springs remained fairly quiet until April 1917, when D.A. McLeod and S.F. Wyckoff opened a new claim, the Horseshoe. Soon they were shipping small amounts of silver-laced copper ore. In 1921 the El Picacho Mining Company reopened the Mexican mine. Mexicans supposedly worked the mine in the 1870s, but verifying this has not been possible. Ore was packed to Reveille for a couple of years. Miners only worked the mines until the early 1920s. The district was dormant until 1938, when the Goldfield Consolidated Mines Company initiated exploratory work, but it found no substantial ore deposits and soon left. The Tamney tungsten property, owned by V.A. Tamney, began to produce tungsten concentrate in 1938. A little more than $9,000 worth of tungsten was extracted before Tamney left in 1940. There also was a small mine at Oak Springs that produced blue chrysocolla, a mineral easily mistaken for turquoise. The mine owners took advantange of the similarity and sold a few hundred pounds of the pure mineral as turquoise. Since 1940 no mining activity has taken place in the district.

The Oak Springs mining district was never short of water. In addition to Oak Springs, there were a number of other springs nearby. A fairly good supply of wood was also available on nearby Oak Springs Butte. The ore mined at Oak Springs had to be transported all the way to Caliente, the nearest shipping point. Because of the cost of getting the ore to the railroad, Oak Springs never really became a good producer. The ore was not rich enough to offset shipping costs and still allow mine owners to make a profit. Oak Springs is located inside the Nellis Air Force Bombing and Gunnery Range, and thus it is not possible to report on the site's present condition.

Petersgold

DIRECTIONS: *From Midway, head south on the old railroad bed for 5 miles to Petersgold.*

Petersgold served the Las Vegas and Tonopah Railroad as a water stop and wood camp from 1907 to 1914. Little was built there besides a water tower and section house. It is obvious from the appearance of the surrounding area that the wood crews must have had to make long trips to find wood. The buildings and water tower were dismantled shortly after the railroad folded. Only one faint foundation marks the obscure site.

Quartz Mountain

DIRECTIONS: *Quartz Mountain is located between Goldfield and Cactus Springs, inside the Nellis Air Force Gunnery and Bombing Range. It is off-limits to the public.*

While initial discoveries made by A.B. Southey, the owner of the Blizzard mine, in 1905 led to the formation of a small camp at Quartz Mountain, it was quickly abandoned, and until 1920 little additional activity took place there. In July Harry Stimler optioned the Southey and Stapin claims for $70,000. But Stimler gave up in 1922, and J.G. Southey and A.L. and Walter Stiles resold the claims for $60,000. The Goldfield Quartz Mountain Mining Company formed and later built a 100-ton mill, the Sailor, in 1926. A number of buildings were constructed to house the work force. The company had limited success and finally folded in the early 1930s. Little else ever took place at Quartz Mountain. By 1950 only a few empty buildings were still standing there. The site is now inside the Nellis Air Force Gunnery and Bombing Range.

Ralston

DIRECTIONS: *From Stonewall, go back the same way you came in and continue going straight for 3 miles to Ralston.*

Ralston, a minor station on the Las Vegas and Tonopah Railroad, was established in September 1907, when the railroad completed its rails through the site. When minor silica mining operations began nearby, a camp of about fifteen formed at the station. A small store and saloon were about all that was ever built at Ralston. After the railroad stopped running in 1914 the nearest railroad was the Bullfrog-Goldfield. The silica ore was then shipped through Cuprite and Bonnie Clare. Ralston's silica production peaked in

1926. More than forty-five carloads of ore were shipped that year. Production in early 1927 was very poor, adding to the troubles of the already faltering Bullfrog-Goldfield Railroad, which relied on the income from the operation. When the railroad stopped running in January 1928 all interest in Ralston faded, and the lonely camp was left to the ghosts. Nothing remains at the site except scattered wood scraps.

Reveille (Gila) Mill

DIRECTIONS: *From Warm Springs, head east on Nevada 375 for 1 mile. Exit right and follow this road for 12½ miles to Reveille Mill.*

The Gila mill, which dates from 1869, was built twelve miles west of Old Reveille because that was the nearest source of consistent water. The main part of the complex was a ten-stamp mill, but there had originally been a smaller five-stamp mill called the Rutland mill at the same site. W.G. Blakely and J.S. French had completed the Rutland mill in April 1867. The Gila mill treated ore from the Mediterranean, the Crescent, and the Atlantic mines at Reveille. The two mills both stopped production in late 1869 when the company owners lost their financial backing.

The district's major disadvantage was the lack of a good water supply. Ore was crushed at night and put through the amalgam process during the day. This process was extremely slow and not very efficient, with a recovery rate of only 35 percent. The ten-stamp mill reopened in 1875 when the Gila

Remains of the Reveille Mill serve as a lonely sentinel over the expanses of Reveille Valley. (Shawn Hall collection, Nevada State Museum)

Silver Mining Company acquired a number of mining properties and needed a place to process its ore. The mill continued to run until 1879. In late 1881 the Continental Silver Mining Company restarted it. By 1883 the mill was running on tailings owned by Governor Jewett Adams, but it was idle again by 1884.

The mill did not reopen until 1904, soon after initial discoveries were made at New Reveille. The mill operated off and on until 1948. Today it is quite dilapidated. A number of small wood buildings that probably housed the mill's employees surround it. A spring at the site, which had a fairly heavy flow, was used to spray the stamps in the mill to keep down the dust. The ruins of the smaller five-stamp mill are only 100 feet from the Gila mill. A small cemetery containing two graves is located nearby. Even though the Gila mill was not actually a town, a description of it is included here because it played an important part in Reveille's history.

Reveille, New

(Morristown)

DIRECTIONS: From Warm Springs, head east on Nevada 375 for 7 miles. Exit right and follow the road for 9 miles. Exit left and follow this road for 12 miles to New Reveille.

New Reveille formed in early 1904, a number of years after Old Reveille had faded into oblivion. The new site was platted directly across the Reveille mountain range from Old Reveille. The camp that sprang up in 1904 was originally called Morristown, after one of the discoverers. Rich lead ore had been located at the site, and soon miners were working a number of mines, including the Last Chance, owned by the Southwest Nevada Mining Company.

The camp had grown enough by the summer of 1904 that a post office, called Morristown, opened on August 29, 1904. C. F. Lupher was postmaster. The initial excitement slowly faded, however, and Morristown slipped into a coma. The post office was renamed New Reveille on June 13, 1905, in the hope that the name change might help the dying camp. But new discoveries were what truly revived it. By March 1906 the Tonopah-Clifford-Reveille stage began running. The fare from Tonopah was $10. The Reveille-Tonopah Mining Company formed. W.W. Booth, publisher of the *Tonopah Bonanza*, was president. At around the same time, Charles Schwab bought out the Southwestern Nevada Company, and the Reveille-Liberty Mining Company became active. By January 1907 New Reveille had a population of 150, and two new companies became active. They were the Bromide Hill Mining Company, of which George Bartlett was president, and the Highland Boy Mining Company, of which Henry Anderson was president. The ore from the mines was sent to the Gila mill, located in Reveille Valley to the west. By 1908 the revival began to slow down. The first blow came in April, when the Reveille-Liberty Company, which owned the Bear, the Silver Buck, the Top Notch, the Liberty Bell, and the Liberty mines, sold its property at public auction because of debts. Herman Rieschke, one of the first arrivals in camp, took over the defunct Highland Boy Company.

The camp continued to decline, and the post office closed on December 31, 1911. In 1913 some new strikes attracted the interest of speculators, but little actual mining took place. By 1925 there were only seven prospectors living in the district. Limited activity took place at the site as recently as 1945 but never lasted long. The New Reveille mines produced just over $30,000 worth of ore, with most of the values being in lead. During the late 1970s and early 1980s the Golden Arrow and Keystone Company, the same company that reworked tailings at Golden Arrow, Keystone, and Old Reveille, reworked the tailings at New Reveille. Most of the buildings left at New Reveille were torn down, but some stone and adobe ruins remain, along with a few mine hoists.

Reveille, Old

DIRECTIONS: From Warm Springs, head east on Nevada 375 for 19½ miles. Exit right and follow the road for 8 miles to Old Reveille.

Old Reveille was one of the first settlements in Nye County. Three prospectors—W. D. Arnold, M. D. Fairchild, and A. Monroe—discovered ore assaying as high as $1,500 per ton there in April 1866, and soon people began to flock to the newly formed district. The camp that sprang up was named Reveille, after the prominent newspaper, the *Reese River Reveille*. By the following summer miners were working more than forty mines in the district, and ore was being shipped to mills at Austin. The two main companies in the district were the Fisherman Mining Company and the Reveille Mining and Milling Company. The Fisherman Company owned the Fisherman, the Victorine, and the Santa Fe mines. The Reveille Company ran the Gila, the Crescent, the August, the Mediterranean, the Atlantic, the National, the Antarctic, and the Adriatic mines. The companies employed a total of seventy-five people and were shipping 200 tons a day. After the Reveille (Gila) mill was built in 1869 some twelve miles to the west in Reveille Valley, ore from the Reveille mines was shipped there. This greatly decreased shipping costs. During Reveille's prominent years prospectors made more

Beautiful stone ruin at Old Reveille. (Shawn Hall collection, Nevada State Museum)

than 950 claims in the hills around town. There were a number of important mines, of which the best producers were the Gila, the Joliet, the Good Hope, the Liberty, and the Fisherman. The Gila was the largest of these, with a 500-foot-deep main shaft and a 1,000-foot tunnel. The richest of the mines was the Gila, which produced more than $500,000 worth of gold and silver before closing in 1891. The small town had a population of 150. Business establishments included a boardinghouse, two stores, and a blacksmith shop. A post office, with John Ernst as postmaster, opened on September 24, 1867, and remained in operation until December 14, 1868. It reopened on July 19, 1870, and remained open until August 21, 1880, except for a short period in the summer of 1875, during which it closed. From 1870 to 1880 Reveille experienced many ups and downs. During the spring of 1871 the town was practically empty. By fall about forty people had returned. By 1874 the mines were idle again.

There were a number of springs and wells in the Reveille area, but their flow was never able to provide the town with sufficient water. Wood was also scarce and had to be brought in by freight lines from Eureka, almost 130 miles away, for use in constructing buildings. All of Reveille's mines closed in 1880, and the town quickly faded. That year Thompson and West report that the town still had thirty residents and contained a hotel, a butcher shop, and a livery stable, in addition to saloons and boardinghouses from earlier development.

Reveille received a new lease on life when the post office went back into business on July 12, 1882. It continued to operate until April 30, 1902.

Reveille never again reached its former level, but it still managed to survive. Even after the post office closed for good, a number of people remained in the town. During the mid-1880s Reveille experienced a revival. The Gila Consolidated Mining Company gained control of the mines and also restarted the mill. The October 1886 shipment from the mill alone was worth $9,200. In July John Leahy, who had resided in Reveille since 1866, died while still working his Daniel Webster and Lady Washington mines. During this period A. F. Spindels was running his Belmont-Tybo-Reveille stage line. Violence came to Reveille in November 1886. Pat O'Brien, a well-known resident of Nye County, attacked Deputy Sheriff J.J. Gallagher. John Reid took away O'Brien's gun, but O'Brien stabbed him badly with a Bowie knife. Reid managed to shoot and kill O'Brien. The death was quickly ruled justifiable. In March 1894 postmaster J.W. Wilsey, who also worked in the mines, was seriously injured when an unexploded charge went off. He survived, but he lost both eyes, a couple of fingers, and his left arm, and his face was lacerated beyond recognition.

When discoveries were made at New Reveille in 1904 a few people chose to live in Old Reveille. As recently as 1911 the town still had more than twenty residents. But after the mines closed at New Reveille, Old Reveille quickly faded. Total production value for the Reveille district, including the old and new towns, was a little more than $4 million.

During the late 1970s and early 1980s a cyanide leaching operation was active at the Old Reveille site. While production levels were respectable, nobody actually moved back to the Reveille district. During this author's visit to Reveille in 1979 the operation was already winding down. The watchman of the site, Bob Walk, graciously provided a grand tour of the operation and of the town of Reveille. At the time, the wood schoolhouse was the only building left, and the company was using it for storage. However, that building has now vanished. All the remaining vestiges of Reveille are stone. An overview of the site reveals the town's layout. A small church and cemetery were on the hill to the west of the ruins, but that site has vanished except for faint raised mounds. Reveille is always cool, a relief after driving through hot and dusty Railroad Valley. The site is well worth the trip.

San Carlos

DIRECTIONS: From Bonnie Clare, go back to Nevada 267. Take a left (east) and follow the road for 2¼ miles. Exit left and follow the road for ½ mile. Take the right fork and follow this road for 2 miles to San Carlos.

San Carlos served as a water stop on the Bullfrog-Goldfield Railroad. It was established in October 1906 and was used until the railroad folded in January 1928. The Las Vegas and Tonopah Railroad passed just west of San Carlos but had its water stop at Wagner. San Carlos never amounted to anything more than a water stop, and even this vanished as soon as the Bullfrog-Goldfield folded. Nothing at all remains at the site.

Silver Bow

(Stephanite) (Wheaton)

DIRECTIONS: From Tonopah, head east on US 6 for 32 miles. Exit right and follow the road for 3 miles. Bear right and continue for 4 miles. Exit left and follow the road for another 4 miles. Exit right and follow this road for 3½ miles. Exit left and follow this road for 8 miles to Silver Bow.

Prospectors made initial silver discoveries in the Silver Bow district in November 1904. Additional discoveries in 1905 and George Wingfield's presence in the district led to a rush to the area. The camp soon became home to more than 300 people. George Wingfield and George Nixon platted the Silver Bow townsite in early 1905, and soon the town became a supply center for the district. A stage line to Tonopah was set up, and the fare was $7. A post office, with Frank Evans as postmaster, opened on September 27, 1905, and the town took on an air of permanence.

Silver Bow had an ongoing problem with claim jumping. One of the better-known incidents involved Edward Johnson and Hugh Fulton, the best of friends. Their heated argument about a claim that neither owned took them to Silver Bow's main street, where Johnson shot and killed Fulton. The death was later ruled self-defense. Claim jumping kept Tonopah's deputy sheriffs traveling back and forth from Tonopah to Silver Bow to arrest the offenders. But these incidents did not deter Silver Bow's growth. By the fall of 1905 the town had grocery stores, saloons, and a few general merchandise stores.

The *Silver Bow Standard* began publication in September 1905. In an attempt to focus attention on Silver Bow, editor Leslie Smaill printed the headline of the January 6, 1906, edition of the paper with ink that had been mixed with gold assaying at $80,000 a ton. Smaill had planned to print the whole paper using the special ink, but, as he wrote in his paper, he realized that the cost of printing an entire edition this way "would amount to a sum far in excess of the real value of the entire plant, editor, force and all." The purpose of the paper was to show that the ore in Silver Bow was no fluke, as the find on the nearby Golden Arrow property was feared to be. The gold used in the paper headline came from the Reed and Robbe claim, located half a mile north of Silver Bow. In 1906 S.A. Knapp and L.V. Cirac heavily promoted a revival townsite, called Stephanite, located only one-fourth of a mile away, but the town began to fail after only a couple of businesses had opened. Businesses in Silver Bow in early 1906 included the Frank Evans General Merchandise; the Silver Bow Trading Company, owned by O.W. Robertson and J.F. Sullivan; the Turf Saloon, owned by Powers and Francis; the Nevada Cash Store, owned by Plamenaz and Company; the Silver Bow Cafe, owned by Brownstead and Whitman; the Silver Bow Club, owned by Derrey and Aubrey; the Overland Cafe, owned by F.W. Butt; the De Sautel Grocery; the Stephanite Club Saloon, owned by Kerran and McCarthy; and the Midway Merchandise Store. In 1906 Dan Fitzpatrick was named mayor, William Enger was selected as justice of the peace, and Hugh Fulton served as constable and deputy sheriff. Frank Evans also built a new road to Tonopah, and P.J. Donohue constructed a five-stamp mill for the Nevada Development Company's property in Haws Canyon. The mill was located to the east, where the water supply was more consistent.

In the Silver Bow district silver was clearly dominant over gold. Most ore contained an average of $1 worth of gold for every $3 worth of silver, but in Silver Bow some claims had as much as twenty times as much silver as they had gold. In 1906 Silver Bow began a quick downhill slide. Most of the mines had panned out, and only a couple of the larger ones remained open. By the next year, fewer than thirty people were in the district, and only the Silver Bow Belle Mining Company continued operations. The post office closed on November 30, 1907. By the end of 1908 the district had been abandoned. The Catlin Silver Bow Mining Company, of which John Gregovich was president, made a new strike in April 1909. It struck ore valued at $100 a ton, but the pocket turned out to be shallow, and it faded before the town could revive. In October the Silver Bow Mining and Milling Company, of which J. G. Crumley was president, organized and worked fourteen claims, including the Black Bear mine. This operation brought people back, and Silver Bow was active enough to be included as a stop on the planned Ely-Goldfield railroad. But after producing little and incurring considerable expenses, the company gave up in 1911. By the beginning of 1912 only the Catlin Company, which employed three, was active in Silver Bow.

A small-scale revival in 1913 warranted the erection of a stamp mill and brought twenty-five people back to the district. William Stevenson moved the old ten-stamp mill from Haws Canyon and rebuilt it at Silver Bow. In January 1913 W. G. Cook shipped ten tons of ore from the Silver Bow Belle — the first ore shipment to come out of Silver Bow in many years. However, only occasional shipments were made during the rest of the year. The revival died before the beginning of 1914, and once again the site belonged to the ghosts. By 1917 only Fred Newton was left, hopefully working on his mine. The last revival took place in the 1920s. The Blue Horse Mining Company, organized in late 1920, built a twenty-ton Gibson mill and sank two shafts. Neither turned out to be worthwhile, and the mill closed in 1921. Thomas Clifford, Fred Jackson, and Frank McMullen worked a number of claims in the 1920s but had only limited success.

The Virgin Gold Mining Company, of which T. L. Mahoney was president, was also active in the district but was never successful. In 1928 the newly formed Silver Hoard Mining Company, which owned four claims in the district, bought its holdings. Silver Hoard dug a number of shallow shafts and short tunnels and contemplated building a fifty-ton mill. But the ore faded before mill construction began.

The last company to work the Silver Bow area was the Silver Bow Consolidated Mining Company, which formed in early 1929. The company purchased the old Blue Horse Company's holdings. The richest mine was the old Blue Horse mine, which was 150 feet deep and had 1,500 feet of lateral work. A fifty-ton flotation mill was built in 1929. It ran sixteen hours a day and employed eleven men. The ore taken from the claims and processed in the mill assayed at $15 to $20 a ton. As was typical for the Silver Bow area, the ore deposits were shallow. By 1930 activity in the district had ceased, this time for good.

The remains of Silver Bow are not very extensive. A few wood cabins still stand and appear to be from the later revivals in the 1920s. Stone ruins and mill foundations also remain. A small cemetery is located nearby. There are mine hoists from some of the mines. Silver Bow is quite hard to reach, for the road is extremely sandy and treacherous. A four-wheel-drive vehicle is recommended.

Stonewall
(Fork's Station)

DIRECTIONS: *From Beatty, head north on US 95 for 53½ miles. Exit right and follow the road for 1 mile. Continue straight for 3 miles. Exit right and follow this road for 1½ miles to Stonewall.*

Prospectors made a small silver strike on the north side of Stonewall Mountain in August 1904. By the end of 1904 there were 150 people living in the camp. The town and the mountain were named for General Stonewall Jackson, who is famous for his stonewall stand at the Civil War battle of Bull Run. The Las Vegas and Tonopah Railroad set up a station nearby. It was called Fork's Station at first but later was listed on timetables as Stonewall. The initial burst of activity quickly subsided. By the end of 1905 only a handful of inhabitants were left. A two-stamp mill was built but wasn't used much.

Several small ore shipments were made after the short boom ended, but the district was essentially abandoned until the Yellow Tiger Consolidated Mining Company moved into the area in 1917. The company was a consolidation of the Desert Chief Consolidated Mining Company, the New Goldfield Sierra Mining Company, and the Red Lion Consolidated Mines Company. The new company was originally known as the Yellow Tiger Mining Company, but the name was altered after reorganization and incorporation.

The company had sixteen claims covering almost 280 acres in the Stonewall district. These claims were in two groups, the Stonewall and the Sterlag. The mainstay of the Stonewall group was the Stonewall mine, which reached a depth of 500 feet before closing. The shaft followed a rhyolite and lime-porphyry vein that varied in width from seven feet to twenty-two feet. The ore assayed at an average of $20 per ton. The gold- and silver-bearing vein was intersected by the Sterlag tunnel, which was 1,100 feet long. The ore from the tunnel contained an average of sixteen ounces of silver and two ounces of gold per ton. The company ceased to operate in September 1920, when its financial backing failed. It was back in business a year later, and

by 1925 the Sterlag tunnel had been extended to more than 5,000 feet. The company left the district in May 1926 to concentrate efforts on its property in the Goldfield district. During the 1920s some other activity took place in the district. In June 1920 Congressman Evans and Ellsworth Oldt discovered a hematite deposit, which they sold to the Red Indian Metallic Paint Company. The company developed and mined the ore for a number of years. The Stonewall area has been dead ever since the hematite deposit was mined out.

The only remains at Stonewall are the mine ruins at the site of the Sterlag tunnel. Nothing else is left. Stonewall is disappointingly bland, especially considering that the ride to the camp is extremely rough.

Sulphide

DIRECTIONS: *Sulphide is located 3 miles north of Jamestown, inside the Nellis Air Force Bombing and Gunnery Range. It is off-limits to the public.*

Sulphide sprang up during the early 1900s and was one of a group of mining camps that included Jamestown, Trappman's Camp, Wilson's Camp, Wellington, Kawich, and Gold Crater. Sulphide was the smallest of these and the first to fold. Its population peaked at twenty-five before excitement in the district faded. Since the camp did not last long or grow to any extent, no substantial buildings were erected. In 1910 Sulphide was still

　　　Preserving the Glory Days

on highway maps, but by 1912 it was gone for good. Because of its location, a present-day status report cannot be made. However, it is doubtful that anything of the small, short-lived camp remains.

Tolicha

DIRECTIONS: Tolicha is located 4 miles southeast of Monte Cristo Springs, inside the Nellis Air Force Bombing and Gunnery Range. It is off-limits to the public.

Tolicha was one of the many camps that formed in the Pahute Mesa area just after the turn of the century. Tolicha was prospected during 1905, but the district remained fairly quiet until July 1917, when three men— Jack Jordan, Ed "Jumbo" Yeiser, and Ed Harney—found rich gold and silver ore on the northern slope of Tolicha Peak. Zeb Kendall purchased one of the larger claim groups, the Life Preserver group, from Jordan and Yeiser for $100,000, and George Wingfield leased it for a while. After Wingfield left, Eric Harvey took over the lease and installed a small Gibson mill. A camp formed, and twenty-five men were employed in the local mines. By September there were forty people living in the camp, and water was hauled in from Monte Cristo Springs. Jordan and Yeiser sold the Landmark group to F.B. Caldwell and R.J. Highland for $20,000. But by 1920 the property had all reverted back to Jordan and Yeiser, and they once again sold their holdings, this time to Thomas Hanley for $225,000. In July a ten-ton mill was completed at Monte Cristo Springs, but it was unsuccessful. Because of the flow of ore, the Tonopah and Tidewater Railroad built a special siding at Ancram.

Harvey continued to work the lease at a profitable level until 1923, when the newly formed Landmark Group purchased the property. In January 1923 Charles Knox, the group's president, purchased the Life Preserver claims from Yeiser for $100,000. He died soon afterwards, and the property was transferred to James Gerard. Gerard sank a 150-foot shaft, which was equipped with a hoist from a mine in Hawthorne. The vein followed by the mine contained mainly gold with traces of silver and produced ore that assayed at $25 per ton. The first payment of $10,000 was made in June. Other active mines in the district were the Bunker Hill, owned by Nick Abelman, Ed Ashton, and Jack Jordan; the Landmark Extension, owned be Dave Llewellyn; and the Periscope, owned by Pat McAuliffe. A short tunnel was dug, but it proved unprofitable. Miners continued to work the Landmark group until 1926, when J.A. Logan and H.L. Gilbert of Tonopah purchased the property for $50,000. Their results were disappointing, and they gave up in 1927, leaving the district totally abandoned.

Charles Mayer, Murray Scott, and Frank Corley leased the Landmark claims during the 1930s, bringing Tolicha back to life. The mine was re-opened and worked until the 1940s, and ore was shipped to the Desert mill at Millers in Esmeralda County. The Landmark mine turned out to be a profit-able investment, producing $180,000 worth of ore during its fourteen years of activity. Mayer, Scott, and Corely were followed by various other leasers, including Charles Whittenburg, Ivy Southey, Fred Schultz, Nick Abelman, Marina Floyd, and Jumbo Yeiser. The workings of the mine were eventually dismantled for salvage. Just before the site was incorporated into the bombing range, a visitor reported that one building remained. Only a few structures ever went up in Tolicha, one of which was the ubiquitous saloon.

Trappman's Camp
(Carr's Camp) (Yellowgold)

DIRECTIONS: *Trappman's Camp is located 4 miles east of Mount Helen inside the Nellis Air Force Bombing and Gunnery Range. It is off-limits to the public.*

Trappman's Camp was a mining camp that formed on Pahute Mesa after Hermann Trappman and John Gabbard discovered silver and gold ore in June 1904. A small camp of about fifteen men soon formed around the discovery site. Wood and water came from Antelope Springs, nine miles away. The newly formed Trappman Mining Company, which was based in Goldfield, moved into the district in July 1905. During the course of the next month, the company's five employees sank a fifty-foot shaft. Ore from the shaft contained an ounce of gold and four ounces of silver per ton. The company continued to work the shaft in hopes of finding richer deposits. When it found none it folded in late 1905.

No one but a few persistent prospectors remained in the district, and even they had left by the summer of 1906. Only Trappman was still prospecting the area. His mine was bonded in April 1908 for $25,000 but quickly reverted back to him when the leasers found little of value. By 1911 even Trappman had left. It wasn't until July 1931 that interest returned. The Yellowgold Min-ing and Milling Syndicate began working on the Yellowgold mine. A small camp formed on the old site of Trappman's Camp and was named Yellowgold. In November John "Curley" Carr and Hugh Shamberger bought the property for $100,000, and the settlement became known as Carr's Camp. Carr began extensive development of the mine and organized the Original Yellowgold Mining Company. For many years just Carr and a couple of workers lived at the camp, and they made only occasional ore shipments. Finally in April 1945 Carr hit the rich vein he had been seeking for fourteen years. Nevada Gold, of which Fry Halloran was president, bought the property in September and named Carr vice president, but he lost interest in the project and left. Carr then organized the Yellowgold Consolidated Mining Company in July 1947. But when the bombing range formed he was forced to leave. In the late 1930s another short-lived camp, which H. B. Kleinstick owned, formed about one and a half miles north of Carr's Camp, around the Wyoming-Scorpion mine. A couple of cabins housed a small work force, but the ore only assayed at $5 per ton, and Kleinstick gave up in the early 1940s. Both sites are now inside the Nellis Air Force Bombing and Gunnery Range and are off-limits to the public.

Wagner

DIRECTIONS: *Wagner is located between San Carlos and Stonewall.*

Wagner served as a water stop for the Las Vegas and Tonopah Railroad from 1906 to 1918. Only two small buildings and a water tower were built, and everything was removed when the rails were torn up after the railroad folded. Only some scattered concrete marks the site.

Wellington

(O'Brien's)

DIRECTIONS: *Wellington is located 6 miles northeast of Gold Crater, inside the Nellis Air Force Bombing and Gunnery Range. It is off-limits to the public.*

Wellington was a short-lived mining camp established in August 1904, shortly after prospectors discovered gold in the nearby hills. The camp of twenty-five only existed for a few months before the ore ran out. A number of buildings and a mill were built at the site. For a while the camp was known as O'Brien's, for one of the first settlers. After the camp was abandoned in early 1905 no other activity ever took place there. As recently as the 1920s a number of buildings and the mill still stood at the site. It is not possible to make a present-day report because of the site's location.

White Rock Spring

DIRECTIONS: White Rock Spring is located 5 miles southwest of Oak Springs, inside the Nellis Air Force Bombing and Gunnery Range. It is off-limits to the public.

White Rock Spring was one of the many small mining camps that sprang up during the early 1900s in southern Nye County. The camp of ten was located at a source of fresh water. Most of the residents worked in mines in nearby Oak Springs. There was one small mine just south of the camp, at Captain Jack Spring. It was never a producer, and when Oak Springs folded, so did White Rock Spring. By 1910 the site was abandoned. No activity has taken place there since. A present-day report on White Rock Spring cannot be made because of its location, but since the camp only lasted for a couple of years, it is doubtful that anything substantial was ever built there.

Wilson's Camp

DIRECTIONS: Wilson's Camp is located ten miles east of Wellington, inside the Nellis Air Force Bombing and Gunnery Range. It is off-limits to the public.

The small mining settlement of Wilson's Camp started after prospectors discovered fairly rich gold and silver ore on the north slope of O'Donnell Mountain. The discoveries were made in May 1904, and soon a camp of about twenty-five formed. A number of short tunnels and shallow shafts were dug, and the ore removed from these mines assayed from $110 to $180 per ton. The mines were on the Pittsburg claim group, owned by O.K. Reed and Ed Slavin. The largest mine was more than 300 feet deep. The ore had an average of six times more silver than it did gold. Although the ore was rich, shipping it out was a problem. The roads to and from Wilson's Camp were extremely primitive, and the ore had to be sacked and brought out by horse or sometimes by a daring buckboard driver.

This obstacle hindered the camp's development. By July 1905 only five miners remained to work the mines. The camp did not become a complete ghost town until late 1906, when everyone gave up hope. Because of the camp's location and the hard trek required to bring in supplies, only one wood building was ever constructed there. This was a fairly small, one-story boardinghouse, in which a number of miners resided. Others erected small stone shelters in which to live. Ten years after activity ceased in the Wilson district, not a sign of the small camp remained.

Southern Nye County

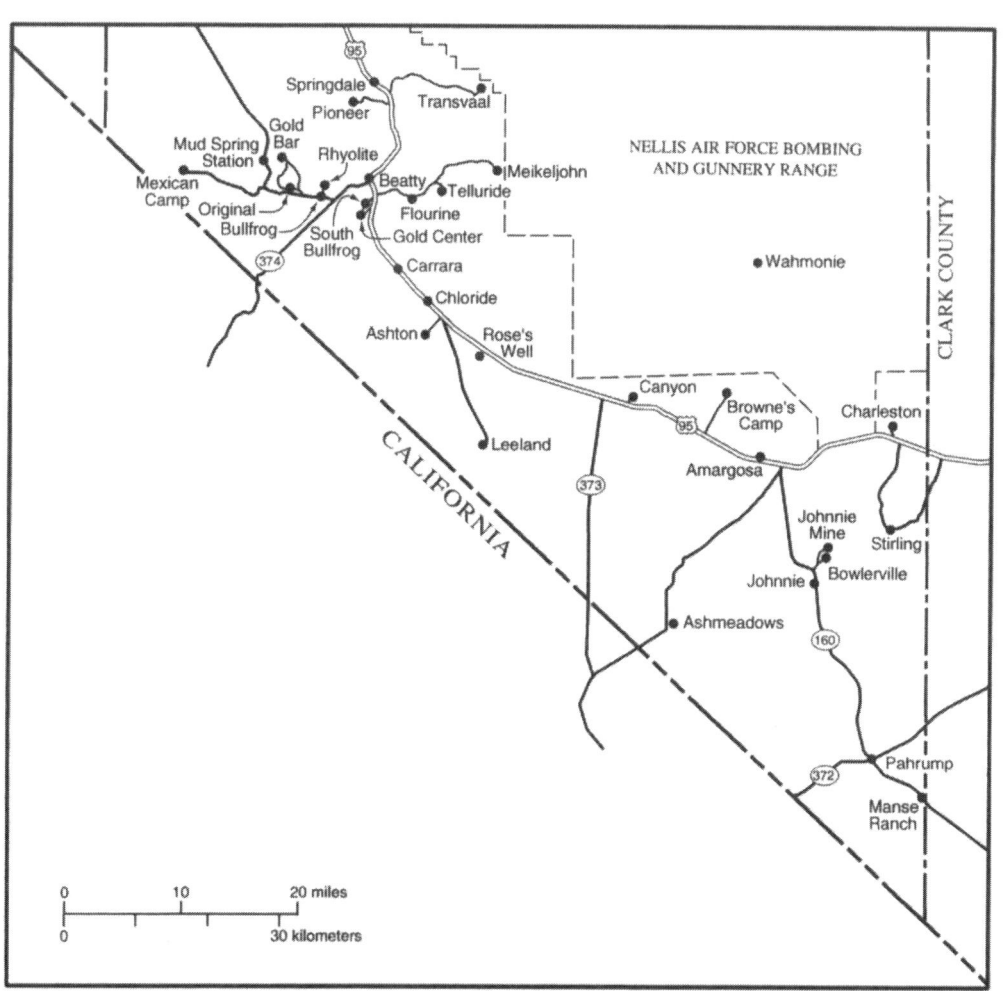

Springdale
Pioneer
Transvaal
Gold Bar
Mud Spring Station
Rhyolite
Mexican Camp
Beatty
Meikeljohn
Telluride
Original Bullfrog
Flourine
South Bullfrog
Gold Center
Carrara
Chloride
Ashton
Rose's Well
Leeland

NELLIS AIR FORCE BOMBING AND GUNNERY RANGE

CLARK COUNTY

Wahmonie

Canyon
Browne's Camp
Charleston
Amargosa
Johnnie Mine
Stirling
Johnnie
Bowlerville
Ashmeadows
Pahrump
Manse Ranch

CALIFORNIA

0   10   20 miles
0   30 kilometers

Amargosa
(Johnnie Station)

DIRECTIONS: *From Amargosa Valley, head east on US 95 for 16½ miles to Nevada 160. Amargosa is ¼ mile west of this point on the north side of US 95.*

Amargosa was a fairly important stop on the Las Vegas and Tonopah Railroad. It was originally a temporary site used while tracks were being laid westward toward Rhyolite and Beatty. The railroad set up the siding in the early 1900s, and it became permanent when strikes at Johnnie created the need for the station. The siding was called Johnnie Station until late 1901, when a post office, named Amargosa, Spanish for "bitter water," opened on December 14. Albert Howell was postmaster. While drilling a well at the site the drillers encountered brackish water, hence the name. The post office closed in November 1902. Another Amargosa post office opened three miles west of Rhyolite in 1904.

By 1904 there were a few buildings at Amargosa. These included a store, a hotel, a restaurant, and a blacksmith shop. The small town remained a pivotal shipping point for Johnnie and other areas to the west. A daily stage to Johnnie was set up, and Amargosa became a vital supply depot for that town. It also became a diversion point for people heading to the new copper discoveries at Greenwater, California, because it was the closest railhead to the booming town until the Tonopah and Tidewater Railroad was completed.

Alkali Bill Brong set up an auto stage from Amargosa to Greenwater, charg-

Depot at Amargosa, 1906. (Central Nevada Historical Society)

Preserving the Glory Days

ing anywhere from $100 to $200 for the seventy-mile trip. In spite of the steep price, Brong was never short of passengers who wanted to experience the thrill of being driven across the desert at fifty miles per hour. All supplies for the booming town were unloaded at Amargosa and then transferred to stages and mule-drawn wagons for the long trek to Greenwater. In 1907, when the Tonopah and Tidewater Railroad was finished, Amargosa ceased to be the scene of much shipping activity. By June 1910 only the railroad agent and a merchant named Rathbone were left in the town.

The town of Johnnie soon became Amargosa's main interest once more, and activity there brought some people back to Amargosa. When Johnnie began to decline again in 1912, Amargosa faded quickly. By the beginning of 1914 Amargosa's population was down to twenty-five. By 1915 the small town was completely abandoned. Today the only remnant of Amargosa is the large concrete foundation of the old station. The site is unrewarding and extremely difficult to locate. The best visitors to Amargosa can hope for is a sense of accomplishment at having found the elusive foundation.

Ashmeadows

(Clay Camp) (Fairbanks Ranch)

DIRECTIONS: *From Pahrump, take Nevada 372 west for 7 miles. Exit right and follow the road for 20 miles to Ashmeadows.*

Ashmeadows, which formed in the early 1900s, was a stop for stages heading to the strikes in the Bullfrog district. A few people had settled in the area before Ashmeadows came into existence. The first was Charles King, who brought 1,300 head of cattle with him and built a stone house at Point of Rocks Spring. Legendary Nye County resident Jack Longstreet lived in Ashmeadows and married a local Indian, Fannie Black. The Lawrence Kimball stages from Indian Springs served the small town. "Dad" Fairbanks, originally from Massachusetts, ran a ranch and was responsible for organizing a small tent city at the fertile site. The tent establishments included hotels, restaurants, and a saloon. When the boom at Greenwater began in the mid-1900s Ashmeadows was the main stop on Fairbanks' Amargosa–Greenwater stage. In 1908 he moved his store from Ashmeadows to Greenwater and in 1910 moved it again, this time to Shoshone. After the booms ended Fairbanks returned to Ashmeadows.

Around 1910 prospectors discovered abundant clay deposits located a few miles to the south of Ashmeadows. These deposits saw little development until the early teens, however. Fairbanks discovered a large deposit in 1916, and commercial production began. The Death Valley Clay Company purchased most of the clay claims and began to process some ore. The mill the company constructed was very inefficient and was plagued with constant problems.

The Death Valley and the Tonopah and Tidewater Railroads eventually built a narrow-gauge spur line to the clay pits at Ashmeadows. The four-mile line had both Plymouth and Milwaukee locomotives and was directly connected with the Tonopah and Tidewater at the Bradford siding in California. Construction of the railroad spur allowed Ashmeadows to continue growing. In 1901 a wooden false-front store was moved from Zabriskie, California, to Ashmeadows. By 1923 the town's population had stabilized at fifty and was the site of a number of substantial buildings, including a few saloons and hotels. The Death Valley and Tonopah and Tidewater Railroads officially folded in 1927 but ran sporadically until 1940. At the clay mine a separate town, called Clay Camp, developed. By 1925 another fifty people were living there. The camp had a school, a grocery store, a saloon, a recreation hall, and a red-light district. The best years for the mine were 1927–1929, when it produced more than $1 million worth of ore. But production levels began to fall off in the 1930s as the Great Depression caused prices to drop. Men were laid off, and Clay Camp slowly emptied. In 1940 the folding of the railroad, which relied on clay shipments to stay in business, put an end to the camp, even though sporadic work continued through 1948. The value of the camp's total production through 1948 was $2.9 million. Some activity has continued at Ashmeadows since 1948. L.A. Chemical reopened the mine on a limited basis and built a mill, but it shut down in 1952 and dismantled the facility. Tenneco, which also built a large mill, reopened the clay pits in the 1970s. In 1976 it sold the operation to the American Borate Company, which built a camp of 100 mobile homes. The mine and mill employed more than 100 men. However, cheap imports ruined the company's plans for the future, and the operation closed in April 1986. The IMV Division of the Floridin Company is currently working the clay mines and employs more than forty people. While both towns have basically disappeared, a number of the buildings were moved to some of the nearby ranches, where they still stand. In 1984 the Ash Meadows National Wildlife Refuge was established nearby to protect the unique pupfish, which lives in the springs throughout Ashmeadows.

Ashton

DIRECTIONS: From Beatty, take US 95 south for 12 miles. Make a sharp right at the exit and follow this road for 2 miles. Exit left on a faint road, the old Tonopah and Tidewater Railroad bed. Follow it for 1½ miles to Ashton.

Ashton, a small water stop on the Tonopah and Tidewater Railroad, was never very important, and no development ever took place there. Only a water tank and a small wood shack were built at Ashton, and neither remains today. After the Tonopah and Tidewater Railroad pulled up its rails, not even trains could visit the parched site. The location is indistinguishable from the desert landscape, and the above directions are based on old railroad maps.

Beatty

DIRECTIONS: Beatty is located at the junction of Interstate 95 and Nevada 374.

The Paiute Indians had been camping at the well-watered site of what would eventually become the Beatty townsite long before white men first occupied it in 1870. A man named Landers built a small ranch there and ran it until the 1890s. Montillus Murray Beatty, also known as Jim, moved into the abandoned ranch house in March 1896. Beatty, a native of Iowa who had fought for the Union in the Civil War, married a Paiute woman shortly before moving to Oasis Valley. His ranch soon became a stop for many tired and thirsty travelers. He and his wife occasionally traveled across Death Valley to stay at their other ranch on Cow Creek.

A small settlement began to form around the ranch in the fall of 1904. As nearby Rhyolite and Bullfrog grew, so did Beatty. In November 1904 Bob Montgomery, who had been involved with mining interests at Johnnie early on, came to Beatty and staked out a townsite. A post office, with Jim Beatty as postmaster, opened on January 19, 1905. The town had no mines but served as a supply depot for the booming Bullfrog district. In contrast to other nearby towns, Beatty had an abundance of water, which came both from springs and from wells. In October 1905 Bob Montgomery completed his $25,000 Montgomery Hotel, which became a landmark in Beatty. In March 1906 Beatty resigned as postmaster and later that year sold his ranch, along with its many springs, to the Bullfrog Power and Light Company for $10,000. He and his wife moved into the town and resided there in a luxurious house.

From 1906 to 1907 Beatty's population reached a peak of 1,000. In March 1907 Montgomery, Charles Schwab, and Malcolm MacDonald sold the Mont-

Beatty, November 1905.
(Nevada Historical
Society)

gomery Hotel and the Beatty townsite to W.S. Phillips and E.S. Hoyt. A number of business establishments, including a Porter Brothers store, opened. Beatty's first railroad arrived in town on October 18, 1906. Four days later the town conducted its first Railroad Day celebration. The Las Vegas and Tonopah Railroad had won the race to Beatty. Within a year two more railroads joined it. On April 25, 1907, the Bullfrog-Goldfield Railroad began regular service to Beatty, and a second Railroad Day celebration was held. The festivities culminated when Old Man Beatty drove a golden spike into the last rail. The last railroad to arrive was Borax Smith's Tonopah and Tidewater, which began service on October 27, 1907.

Beatty had two early newspapers, one of which printed only a few issues. The second, the *Beatty Bullfrog Miner,* was one of the two most important papers in the Bullfrog district. The other was the *Rhyolite Herald.* The *Beatty Bullfrog Miner,* known as the *Bullfrog Miner* during its first month, began publishing in Beatty on April 8, 1905. C.W. Nicklin ran the paper, which competed directly with the *Bullfrog Miner* published in nearby Rhyolite. Clyde Terrell and Dan McKenna bought the paper on January 12, 1907. The two men had some problems and on June 15, 1907, sold the paper back to

Nicklin. He continued to publish the paper but gradually lost interest. On February 22, 1908, he sold the paper back to Clyde Terrell, who ran it until February 1908. Nicklin, much to his dismay, had the paper again. He sold out to Earle Clemens, owner of the *Rhyolite Herald*. He finally administered the paper's last rites in July 1909.

Montgomery's hotel was moved to booming Pioneer in 1908. It burned in that town's fire in May 1909. Jim Beatty, the beloved founder of Beatty, died in 1908. Beatty had been hauling wood when he fell off the wagon and hit his head. He died the next day, on December 14, at the age of seventy-three. The town named for him now reigns as the sole survivor of the once-booming Bullfrog Mining District. Long after Rhyolite and Bullfrog had faded, Beatty continued to serve as a supply depot for people still living in the area. In 1938 the United States Milling Company started a 100-ton mill to treat ore from its operations at Rhyolite, and its presence boosted Beatty's flagging economy. But the onset of World War II forced the plant's closure, and it never reopened. The railroads are all gone. The Las Vegas and Tonopah left in the teens, the Bullfrog-Goldfield left in 1928, and the Tonopah and Tidewater left in 1940. Robert Crandall of the *Goldfield News* published a newspaper, the *Beatty Bulletin,* from April 25, 1947, until December 28, 1956, but it later merged with the *Tonopah Times-Bonanza.*

The recent renewal of large mining operations at Rhyolite has brought new residents to Beatty. Increased tourist traffic has led to the construction of new businesses, including casinos. Jo and Jerry Mundt currently publish a small, typewritten newspaper called the *Beatty Newsbits*. It includes newsy items about the local area. The town is full of history, and there are many interesting remnants from Beatty's early days. Food, gas, and lodging are all available. Beatty remains alive and well.

Bowlerville

DIRECTIONS: *Bowlerville is located 2 miles south of Johnnie.*

Bowlerville was a short-lived mining camp that formed around the Bowler mine shortly after the turn of the century. The small camp was named after Fred Bowler, founder of the town. The population never exceeded fifteen, and the camp only lasted a year and a half before folding completely. Nothing substantial was ever built at the site, and not a trace of the camp remains. Only a small tailing pile from the Bowler mine shows that people were ever here.

Browne's Camp

DIRECTIONS: From Lathrop Wells (now called Amargosa Valley), head east on US 95 for 9 miles. Exit left on a faint dirt road and follow it for 2 miles to Browne's Camp.

Browne's Camp was a short-lived mining camp that formed in 1907 or 1908. A small tent camp sprang up after prospectors discovered some shallow ore deposits. The camp received supplies from the Chloride stop on the Las Vegas and Tonopah Railroad, just a few miles to the southwest. Browne's Camp disappeared from maps in 1920 before any actual production took place. Nothing remains at the site.

Bullfrog

(Bonanza) (Amargosa City)

DIRECTIONS: From Beatty, take Nevada 374 west for 3.9 miles. Exit right and follow the road for 1½ miles. Exit left. Bullfrog is ½ mile down this road.

Both Bullfrog and its sister city, Rhyolite, formed soon after Frank "Shorty" Harris and Ernest L. Cross made initial discoveries in the hills just west of the future Bullfrog townsite on August 4, 1904. There are different versions as to the origin of the name Bullfrog. Harris's version is that *he* discovered the ore, which was speckled green, and thought it looked like a bullfrog. Cross's version, and what seems to be the more widely accepted story, is that the name came from an old song he always sang that included the line, "The bulldog on the bank and the bullfrog in the pool."

The entire tent town of Amargosa City, which was located near the original Bullfrog mine, moved to the Bullfrog townsite in March 1905. In November 1904, however, before the move took place, Carl Stoddard established another town, called Bonanza. When the Amargosa Townsite Company bought the Bonanza townsite, it was renamed Bullfrog. In addition to tents, a small number of stone cabins and dugouts were built. Brush and small trees from the nearby hills provided the only source of fuel for the drafty buildings. In early 1905 Rhyolite began to form, and soon there was fierce competition between the two towns. Founders of Bullfrong offered free lots to merchants and potential residents in March 1905. A post office, with Leonard McGarry as postmaster, opened on March 21, 1905, and served the 300 permanent residents of Bullfrog. A paper, the *Bullfrog Miner,* owned by Frank Mannix, began publication on March 31, 1905. Initially published in Tonopah, the paper was published in Bullfrog until March 23, 1906.

May 1905 was the peak of Bullfrog's short existence. H. H. Clark, who had

Bullfrog, Nevada, November, 1905

earlier given Main Street lots away, was now selling them for as much as $1,500. A $65,000 water system; an $18,000, two-story hotel; and a county jail were all constructed in early 1905. Before the completion of the water system, water had to be brought in from the Amargosa River. Businesses in town included the three-story Merchants Hotel, owned by Casey and Arden; the Denver Lodging House; the McGarry General Store; the McDonald Livery; the Southern Nevada Banking Company; and an icehouse. Saloons included the Amargosa Club, the Bullfrog, and the Combination. Bullfrog was also the terminus for the Goldfield Auto Stage Company. Former Nevada Senator William Stewart, then eighty, moved to Bullfrog in May and built a $20,000 complex that consisted of a one-story, ten-room adobe house and a large law office. Stewart had a 1,200-volume law library in his office; it is believed that the library was one of the largest in the state. He also owned an entire block on Bullfrog's main street and constantly expressed his opinion that Bullfrog would soon outgrow Rhyolite.

Stewart's hopes were crushed as Rhyolite continued to swell, draining more people from the fading Bullfrog. A number of fatal gunfights did not

help Bullfrog. The May 4, 1906, issue of the *Rhyolite Herald* proclaimed: "Verily, the Bullfrog Croaketh." The article said that the last mercantile in Bullfrog had closed and moved to Golden Street in Rhyolite. Soon Bullfrog had only a handful of residents. By May the last business in Bullfrog had left for Rhyolite. The now-vacant hotel, Bullfrog's last status symbol, burned on June 25, 1906. It was a funeral pyre for the dying town.

By 1907 Bullfrog was practically empty. Only Stewart and a few others still hoped for a boom. Stewart's confidence in Bullfrog finally faded, and he moved to Washington in May 1908. With Stewart gone and Rhyolite in the midst of its dramatic decline, Bullfrog's last hopes for survival were dashed. The post office struggled into 1909, but postmaster James Thomas finally closed its doors for good on May 15, 1909. Bullfrog joined the growing ranks of Nye County ghost towns.

Today Bullfrog is almost completely flattened. The only visible remains are those of the old icehouse. These crumbling adobe ruins are not going to last much longer. Soon Bullfrog will have returned to the earth from which it so quickly sprung. Some faint ruins of Senator Stewart's house and office still exist, but they are extremely difficult to locate. His library and office sign are now part of the Nevada Historical Society's collection. For many years Mrs. Earl Gregory ran a small antique and coffee shop in Bullfrog, but the shop closed when she died in the 1980s. If you stop at Bullfrog be sure to visit the Bullfrog-Rhyolite cemetery, located just south of the Bullfrog townsite.

Canyon

DIRECTIONS: *From Amargosa, head west (north) on US 95 for 15 miles. Canyon is located just past the Skeleton Hills.*

Canyon was a small water stop on the Las Vegas and Tonopah Railroad. The railroad used Canyon, which was built in 1906, until the rails were torn up a dozen years later. Nothing of interest remains at the site.

Carrara

(Arista) (Gold Ace) (Hollywood)

DIRECTIONS: *From Beatty, head south on US 95 for 8.2 miles; then exit left. Ruins of Carrara are ¼ mile off the highway.*

Initial activity in the Carrara area in 1904 pointed towards some very promising marble outcroppings, but they turned out to be too fractured to sell. The search continued for additional deposits. In 1911 prospectors found new deposits that were less fractured, and the American Carrara Marble Company formed. P.V. Perkins was president. The company had strong financial backing from stockholders in the East.

The quarry was located in the mountains, and the townsite was platted about three miles away on the flat valley next to the Las Vegas and Tonopah Railroad. Perkins began preparing the quarry for mining operations by removing all the topsoil and installing a sixty-five-foot derrick with a sixty-foot boom. It was able to handle marble blocks weighing up to fifteen tons. Because of the distance from the quarry to the railroad, work on a three-mile spur line began in 1913. Construction went slowly because the desert conditions only allowed the equipment to function properly for two hours in the early morning.

Perkins laid out the Carrara townsite, and a nine-mile water pipeline was built from Carrara to Gold Center, five and a half miles northwest. Perkins also organized the Carrara Townsite Company and planned a Townsite Day to officially dedicate the town. It was held on May 8, 1913. Some buildings were moved to Carrara from Beatty and Rhyolite to make the town look more finished than it really was. Townsite Day was a gala celebration. A band from Goldfield played, and there was a baseball game. A post office, with Frederick Boyberg as postmaster, opened on May 24. The Hotel Carrara, run by Bert LeClair, opened in June. The hotel had originally been the Bonanza Hotel in Rhyolite and featured electric lights, baths with running water, and telephones. Soon a town newspaper, the *Carrara Obelisk,* was founded. Its advertising support came from Goldfield and Beatty. It was published in

Salt Lake City, and all stockholders in the company were subscribers, which boosted circulation.

The railroad to the quarries, which was completed in early 1914, consisted of two flatcars operated by the Lidgerwood cable system, which ran by counterbalance. As the loaded car descended from the quarry, it pulled the empty one up the hill using counterbalance. Both cars ran on the same rail, and a turnout at the middle of the railway enabled them to pass each other. A powerhouse built at the top of the mountain helped protect the equipment from the numerous landslides that occurred during wet weather. Near the Las Vegas and Tonopah Railroad siding, the marble company built huge saw tables to cut the marble into transportable sizes. Although initial quarrying began in 1913, it was not until April 7, 1914, that the company made its first shipment of six huge blocks of blue-white marble to Los Angeles. Soon more shipments had left for other areas. By the summer of 1914 there were twenty-five houses in Carrara. Businesses included the Carrara Hotel, the Hot Potato Dance Hall, Mrs. Dan Lennon's restaurant, Fred Boyberg's general store, and the Exchange Bar. A well-defined street layout was in place. The names of the streets were Verde, Pertelle, Sicilian, Carrara, Pompeii, Sierra, Parian, and Rouge. The Carrara school district was established in January 1915, and Mina Smith was the first teacher. In April a 1,000-pound-a-day ice plant was completed. A new park featured Carrara's most prominent landmark, a large fountain with multicolored lights. This fountain led to the creation of Carrara's unofficial slogan, "Meet me at the fountain."

Carrara's peak years were 1915 and 1916. At that time there were more than forty buildings at the townsite, and there was also another camp at the quarry. Total population was close to 150. A dozen buildings went up at the quarry camp, and the company payroll was $6,000 a month. Marble from Carrara won a gold medal at the Panama-California Exposition. But the town was only a successful marble producer for a short time. The marble tended to be fractured and not pure, and Vermont began turning out large amounts of higher-quality marble. In late 1916 the Nevada-California Power Company cut off all electrical power to the town because it wasn't profitable to provide it. This halted all activity at the quarries. In 1917 the *Carrara Obelisk* ceased publication, putting another nail in Carrara's coffin.

After World War I the Las Vegas and Tonopah Railroad shut down, essentially cutting Carrara off from the rest of the world. A small-scale revival took place in the early 1920s, but the operation soon shut down, and Carrara slipped into oblivion again. The final blow came on September 15, 1924, when the post office closed. Carrara was now one of the many Nye County ghost towns. It took a last gasp in 1928 when prospectors discovered some gold deposits a mile to the north. The rush brought some people back to Carrara, but most lived at a new camp, Arista, that formed near the mines. The *Carrara Miner,* with Bernard Store as editor, began publication in 1929.

It was basically a promotional paper of the Gold Ace Mining Company. The new camp advertised itself in the paper like this: "If you want to hitch your wagon to a star, get located while the going's good, pick out your lot in the new town of Arista." Ray Boggs, Briz Putnam, and Harry Stimler developed the townsite. A Townsite Day was held on June 23, 1929, but there weren't many takers for the lots. The camp's population quickly rose to more than fifty. The Gold Ace was the main mine and was listed on the Los Angeles Curb Exchange. By January 1929 the company had completed a 200-ton mill, and it employed fifty people. There were many other active groups in the district in 1929. The Arista Mining Company employed eighteen and had initial assays of $1,285. Other active groups were the Gold Ace Consolidated; the Indian Mines, Inc.; the Gold Ace Annex; the Gilroy; the Yellow Jacket; and the Gold Ace Extension. However, by the fall of 1929 it had been determined that all the mines were worthless, and everybody had left by end of the year.

Gold Ace stubbornly refused to die. New strikes in February 1930 revived interest in the camp, but production didn't begin until the summer of 1931, when the Beatty Gold Mines and Milling Company, of which Fred Kuenzel was president, took over the mines. Hollywood interests backed the company heavily, and the old camp of Gold Ace, or Arista, was renamed Hollywood. By June the camp had nine houses, an office building, a store, and a boarding-house. Ore values were promising enough that the company added a second shift in October. During the next summer it hired an additional fifty men to build a fifty-ton flotation mill. But the company ran into financial trouble, and the Gibraltar Gold Mines Company took it over in December. The mill, which D. F. Meiklejohn and H. C. Downey built, was completed in January. The Gibraltar Company was active until 1936, when the Mark Requa Estate, which had a lien, bought the property for $112,000. There wasn't any mining activity after the company folded, and total production value for the district is $350,000. Old Carrara saw some activity beginning in 1941, when the Carrara Portland Cement Company built a large plant there. A seventy-one-ton tube mill was moved from Eureka by truck and trailer. Up to that time, it was the largest piece of equipment that had ever been moved by that method. By May the company employed sixty, but because the industry was not considered essential by the War Department, it was forced to close down, and it never reopened.

Nothing substantial remains at Carrara today. As recently as the mid-1960s a few buildings were still partially standing, but now just piles of rubble and faint foundations mark the site. The area is covered with broken glass, china, and thousands of old tin cans. The site can be identified by a solitary chimney standing in the middle of the desert. The road to the quarries is still passable. The buildings just to the north were never part of Carrara but are the remains of the Elizalde Company from the Philippines, which erected all the buildings and then abandoned the site in 1936 before production even began. At Gold Ace, Arista, or Hollywood, foundations and collapsed buildings mark the townsite. Mine dumps, collapsed shafts, and mill foundations are nearby.

Charlestown

DIRECTIONS: *From Amargosa Valley, head east on US 95 for 26½ miles. Exit left and follow the road for 1¾ miles to Charlestown.*

Charlestown was a small and obscure railroad station and mining camp. It came into existence in 1906 when the Las Vegas and Tonopah Railroad established a station as construction progressed westward to Rhyolite. Limited mining occurred at the site in the early 1900s, but after the Las Vegas and Tonopah folded and tore up its rails in 1918 the camp was completely abandoned. Only some wood rubble from the water tank tower marks the site. Just beyond this are the remains of a small mine, which never produced any paying ore.

Chloride

DIRECTIONS: *From Beatty, head south on US 95 for 10 miles to Chloride.*

Chloride was a small water stop on the Las Vegas and Tonopah Railroad. Built in 1906, it remained in use until the rails were torn up in 1918. The stop also served nearby Carrara as both a supply depot and a shipping point for the Carrara Marble Company when the town was at its peak. Nothing of interest remains at the site.

Gold Bar

DIRECTIONS: *From Bullfrog, continue west for 1 mile. Take a right at the fork and continue down this road for ½ mile. Take another right at the fork and follow the road for 3½ miles to Gold Bar.*

Gold Bar was a small mining camp in the rich Bullfrog Mining District. The camp was the most important of those that sprang up in 1905 in the area west of Bullfrog. The Bullfrog Gold Bar Mining Company immediately formed and ran the promising mine. Because of the increasingly large work force, a townsite was laid out in November 1905. The camp received a big boost in 1906, when Charles Schwab bought an interest in the company.

The mainstay of the camp was the Homestake mine, which had ore worth an average of $30 per ton and whose ore was sometimes worth as much as $150 per ton. An impressive ten-stamp mill was built at the site and started up in January 1908. The crushed ore was shipped on the Las Vegas and Tonopah Railroad, located a few miles to the south.

A camp of more than fifty grew up around the mine and mill, and a number of substantial buildings were erected. By early 1908, however, the ore began to mysteriously fade. The mill closed in May 1908, and soon work on the mine also stopped. The mine was over 500 feet deep when operations ceased, and it had almost 5,000 feet of lateral tunnels. The Gold Bar Company officially folded in December, and J.P. Loffus bought the property at auction for $39,000. Loffus was president of the old company, and he formed the New Gold Bar Mining Company. Stockholders in the old company were furious because they had been led to believe that the mine was extremely rich. However, it turned out that the original company had been promoting a fraud. Local papers had claimed that $1 million in gold was in sight. Loffus had been running the company from afar and had been funneling money in to develop the mine and build the mill. Yet, nothing much had been produced. He traveled to the mine and found that the ore was worthless. He had been swindled out of a small fortune before becoming aware of the fraud. He forced the old company out of business and formed the new company to try to recoup his losses. That never happened, and in February 1911 the mill was sold and moved to Round Mountain. Only a few ruins mark the site, because most of the buildings were moved to Rhyolite during its boom. There is quite a bit of scattered rubble in the area, but little else is there. The mill foundation is easy to locate, and the partially collapsed Homestake shaft is just above the mill ruins.

Gold Center

DIRECTIONS: *From Beatty, head south on US 95 for 2½ miles. Exit right and follow the road for ½ mile to Gold Center.*

Gold Center was an important terminus for three railroads during the early 1900s. The townsite, platted in late 1904, was a critical water source for nearby Bullfrog and Rhyolite, both of which were then beginning to flourish. Even though the Amargosa River was barely a few feet wide, it was the only water supply within miles and was highly treasured. The origin of Gold Center's name is a mystery, for when the town formed in 1904 it wasn't the center of anything and had no gold. A post office, with Ben Burger as postmaster, opened on January 21, 1905, but the town was not really on the map until 1906.

The Gold Center Ice and Brewing Company's brewery in 1908. (Nevada Historical Society)

In June 1906 the Las Vegas and Tonopah Railroad reached Gold Center. W. C. Bryan set up a railroad yard that included a number of station buildings, switches, turntables, and a 2,000-foot sidetrack. The yard was located on the 116.1 mile marker for the railroad and was used until the railroad reached Rhyolite. In 1907, when the yards were completed at Rhyolite, those at Gold Center were abandoned. The first freight train pulled into Gold Center on October 7, 1906, with Governor Sparks on board. The first passenger train, loaded to capacity, came just five days later.

The Tonopah and Tidewater Railroad, after much delay, reached Gold Center in late 1906 and eventually continued on to Goldfield. The Bullfrog-Goldfield Railroad also ran through Gold Center. After it was abandoned in 1914 the railroad's right of way was incorporated into the Tonopah and Tidewater Railroad. Gold Center reached its peak in 1907. At that time the town had a post office, a hotel, a bank, brokerage firms, a few mercantile stores, and a score of saloons. A newspaper, the *Gold Center News,* which began publication in September 1906, reached a peak circulation of 100 in early 1907 but folded later that year. The most imposing and unique building in Gold Center was the Gold Center Ice and Brewing Company's combination brewery and icehouse. The company built this structure in 1907 and was soon supplying many of the saloons in Gold Center and Rhyolite. At the time, the company had the only distilled water ice plant in Nevada. An adjacent well supplied the ten-ton ice plant with water. The company did not last too long, however, and the brewery closed in late 1908 or early 1909.

The Gold Center Water and Mills Company did the only mining in the Gold Center area. From 1907 to 1909 the company sank a number of shafts and dug a few tunnels but discovered no worthwhile ore. The company did build a thirty-ton concentration mill to serve the district, but few mining companies wanted to send their ore to Gold Center, and the mill soon closed.

When nearby Rhyolite began to decline in late 1908, Gold Center also faded. The post office closed on November 30, 1910. By the beginning of 1911 the population was down to twenty-five. The town struggled to survive, but when the Las Vegas and Tonopah Railroad tore up its rails in 1918–1919, the end had come. The scant remains of Gold Center consist mainly of the stone foundations of the brewery and mill. Scattered wood boards give a general idea of where the town was located. The harsh desert climate has almost completely erased all signs of the once-bustling railroad town. The nearby Gold Center Ranch formed after the death of Gold Center.

Johnnie
(Montgomery) (Johny)

DIRECTIONS: *From Pahrump, follow Nevada 160 north for 16 miles to Johnnie, located on the west side of the highway.*

Johnnie formed in 1891 soon after prospectors discovered gold at the Johnnie mine site, a few miles northeast of the townsite. (A separate section is devoted to the Johnnie mine and the history of that site. This section deals specifically with Johnnie.) The Congress and the Johnnie mines were the mainstays of Johnnie. The Congress, also known as the Chispa, was just west of the Johnnie townsite and was also discovered in 1891. A small rush, made up mostly of people coming from fading mining camps in the north, soon took place. By May 1891 more than 100 men and women were living at the camp, then known as Montgomery. The town was named for the Montgomery brothers, who had made a number of the initial discoveries there. A post office called Montgomery, with Mason Bartlett as postmaster, opened on August 7, 1891. Dan McDonald brought supplies in via the Santa Fe Railroad from Daggett, 150 miles away. Water for the camp came by donkey from Horseshutem Springs, four miles east across Pahrump Valley, and a log-hauling road to Mount Charleston was also built. The camp contained a hotel, a blacksmith shop, and a store. The discovery of ore worth $3,000 a ton by someone who was scraping the area for the blacksmith shop was the source of great excitement. The shop was then built in a different location. But by 1895 only the Chispa mine was active. The new ten-stamp Chispa mill was completed in September, and seventy men were employed, but the veins turned out to be quite shallow, and the camp quickly emptied. The

Montgomery post office closed on March 17, 1894, and a ghostly silence fell over the camp.

A Utah mining company bought the Congress mine in early 1898, and soon more than fifty people were back at the camp. A post office named Johny, with Samuel Godbe as postmaster, opened on June 28. There were a number of labor disputes at the Congress mine during the next few years, including one in which the leasers and the mine owner had a major disagreement over the ownership of the mine. The dispute was not resolved, and the two men ended up dead. During the dispute, workers burned the mill and a cookhouse, dynamited the mine office, and looted the safe. After the mill burned in December 1898, all operations were shut down. The Johny post office closed on April 18, 1899, and the town was once again a ghost.

Soon after prospectors made discoveries in Goldfield and Bullfrog in 1904, some other prospectors drifted down to the Johnnie area. New discoveries in 1905 prompted the reopening of the Congress mine, and initial work on a number of new claims also began. The post office, called Johnnie, was back in service on May 27, 1905, and remained in operation until December 31, 1914. A new townsite was platted at the old site in May 1905, and the lots sold very quickly. By early 1907 almost 350 people were living in Johnnie. The townspeople used the structures that were left from earlier activity and built several new stores, saloons, hotels, and restaurants. A daily stage to Amargosa formed on the Las Vegas and Tonopah Railroad. The fare was $3. The Johnnie Consolidated Mining Company purchased the Congress mine in 1908 and worked it until 1914. Activity faded in that year, and the town's population dropped from 100 to less than 15. After the post office closed, Johnnie embarked on a downhill slide until new discoveries rescued it again in late 1915.

Prospectors discovered placer gold around the townsite, and soon Johnnie had some of its old vim and vigor back. The post office once again reopened on April 14, 1916, signaling the beginning of Johnnie's last and longest revival. The Eureka Johnnie Gold Mining Company became active in 1921, when it sank four shafts and built a small two-stamp mill. Ore processed by the mill assayed at about $10 per ton, and the company continued to work the district until 1925, when financial woes forced it to fold. Johnnie did not fold, though, for the placer gold operations were rich enough to keep people in town. The post office closed for good on November 6, 1935, but a handful of people remained. New activity in the late 1930s at the Johnnie mine drew most of the people from the town, and the site was soon empty.

Matt Kusic purchased the whole site after World War II and then sold it to Al Padgett, who in turn sold it to Deke Lowe, the present owner. Lowe has revived the camp, not as a mining venture but as a secluded home for a number of people.

There are numerous remains at the site, including a few wood cabins that are slated for restoration. It is always encouraging to see a desolate ghost town being reclaimed and protected from the harsh desert and rampaging vandals.

Johnnie Mine
(Labbe Camp) (Yount Camp)

DIRECTIONS: From Pahrump, head north on Nevada 160 for 17 miles. Exit right and follow the road for 1½ miles to the Johnnie mine.

Five prospectors searching for the lost Breyfogle mine discovered the Johnnie mine in 1891. Many believe that their rich discovery is that lost mine. The mine was named after Ashmeadow Johnnie, an Indian from Pahrump Valley. Ed Oldfield discovered another mine, the Crown Point Globe, in 1891, and he built a one-stamp Kendall mill at the site. A small camp, named Yount Camp after one of the discoverers, developed, and by May ten men were living in tents at the mine.

By the late 1890s the mines were idle. In May 1907 D. G. Doubleday bought the Johnnie mine for $300,000 and immediately hired fifty men. On December 24, 1908, a sixteen-stamp Nissen mill, which cost $100,000 to build, was completed at the Johnnie mine. By the end of 1908 the Crown Point Mining Company was also active. A small camp at their mine consisted of a five-room house, a boardinghouse, a bank, an office, a blacksmith shop, and a storage house. The mine and mill continued to operate until 1914, and production value for the Johnnie mine from 1910 to 1914 was close to $4 million. In April 1915 A. P. Johnson bought the property at a sheriff's sale for $10,000 and reopened the mine, which was eventually dug to a depth of 1,100 feet, with 12,000 feet of lateral workings. The sixteen-stamp mill, which had a capacity of eighty tons, reopened, and the mine and mill employed a total of ten men. Johnson and his associates continued to work the Johnnie mine until 1924.

The Crown Point Globe mine contained rich pockets of gold ore, but the values were erratic. One shipment of 150 tons yielded almost 14,000 ounces of gold. The mine was dug to a depth of 200 feet and had more than 500 feet of drift tunnels. It still retains most of its original machinery, including the old 1891 hoist. The Crown Point Globe mine is also known as the Overfield mine because the Overfield Mining Company of Chicago owned and worked it for a number of years.

The Johnnie mine was the scene of extensive activity beginning in 1937. A regular camp formed, and it included a store, a pool hall, and other businesses. A post office opened on September 14, 1937, and remained open until July 1, 1942. Water for the active camp came from nearby Grapevine Springs

via a four-inch pipeline. The camp was also known as Labbe Camp, after Charles Labbe. Labbe was one of the principal owners of the mine and mill and had been working the district since the 1920s. Production stopped in 1942, and George and Judy Warner later purchased the property. Production value for the district from 1934 to 1942 was in the neighborhood of $25,000, while total production value for the Johnnie district during this period was $1.6 million.

The ruins at the Johnnie mine are fairly extensive. The sixteen-stamp mill is quite well preserved and is ready to use if the mine ever reopens. A number of buildings remain at the Johnnie mine, but the site is private property, so please request permission before exploring. About a mile north of the Johnnie mine is a small group of wood buildings built during the activity of the 1930s. An intriguing mystery surrounds the two graves located near the highway just below the Johnnie mine. Art Davidson, author of *Sometimes Cassidy — the True Story of Butch Cassidy,* claims that they are the graves of Robert Leroy Parker (Butch Cassidy) and his brother, George Cassidy Parker. Many others strongly agree with Davidson. Davidson believes that the two men were not killed in South America, as is commonly thought, and, using research, presents a strong argument for the fact that the two owned a number of mining claims at the Johnnie mine, including the mine itself. He presents evidence suggesting that A. P. Johnson was, in fact, an alias and that the family, which also owned the Florence mine (named after the Parkers' daughter) in Goldfield, ran the Johnson Mining and Milling Company. The Parkers were Mormons and sent much of their profits to Salt Lake City. George died at Johnnie and was buried. Butch died in Leeds, Utah, in 1956, and his family secretly brought the body to Johnnie, where they buried him with George. An interesting tangential note is that the two graves do not lie side by side but rather head to toe. Mormons believe that by positioning graves this way, when the resurrection occurs, the deceased will not be able to see who is called first. Davidson's book is incredibly intriguing, and while his assertions may be difficult to swallow, the evidence is compelling. Whether to believe Davidson or not must be left up to the individual reader, but the two graves are there, just as he describes. Family members of the Parkers still reside in Goldfield. The total production value for the Johnnie mine and Crown Point Globe mine is reported to be as high as $20 million, but extensive research suggests that this figure is exaggerated, and that a more accurate figure is around $10 million. Still, the mines' production record is impressive. Be sure to pack plenty of water when visiting the Johnnie mine, because the Pahrump Valley area tends to be extremely hot and dry.

Leeland

DIRECTIONS: From Beatty, head south on US 95 for 12 miles. Exit right and follow the road for 9 miles. At the fork, exit right and follow the road for 1½ miles to Leeland.

Leeland, a small railroad station on the Tonopah and Tidewater Railroad, was established in 1906 at the 144-mile marker. The station became an important shipping point for Lee, California, which boomed five miles to the west in 1906. Regular train service through Leeland began on October 15, 1907. A week later the three-room station and a Wells Fargo office were completed. They had dirt floors and no plumbing or electricity. C. E. Johnson started a stage line to Lee. As Lee continued to grow, so did Leeland. By 1911 twenty-five people were living at the station.

A post office, with George Railton as postmaster, opened at Leeland on November 23, 1911, and the town reached its peak the following year. After Lee folded, Leeland began to fade. The post office closed on November 14, 1914, and the town slowly sank into oblivion, although the railroad still used the site as a water stop. The small depot remained in existence until April 1931, when a demented prospector, Jack Behresin, burned the station house and other facilities and then committed suicide. An experimental farm began operating nearby in 1915 and gradually became known as the T and T Ranch. The Leeland Water and Land Company offered homesteads, but the only takers were officials of the Pacific Coast Borax Company, which had nearby operations. Gordon and Billie Bettles ran the ranch from 1946 to 1964. Since then, the ranch has slowly been parceled off. The Tonopah and Tidewater tore up its rails in the early 1940s, leaving Leeland to the ghosts.

Absolutely nothing remains at Leeland. The only way to locate it is to follow the old railroad right of way to a point where the road to Lee used to run. The station site was here. The lack of remains, along with the difficulty of locating the remote site, make visiting Leeland more trouble than it is worth.

Manse Ranch
(White's Ranch) (Yount's Ranch)

DIRECTIONS: *From Pahrump, head south on Nevada 160 for 6½ miles to Manse Ranch.*

Mormon Charlie, a Paiute Indian, founded Manse Ranch in the 1860s. In 1877 Joseph Yount and his family bought the ranch after Indians had killed their horses, forcing them to settle down. There was a large spring at the ranch, and soon Yount had a 160-acre spread. In addition to cultivating the famous orchard, Yount also raised chickens and milk cows. The Manse Ranch, also known as Yount's Ranch, was the only stopping place for prospectors traveling from Las Vegas to mining camps in the west. Everyone was welcome, and all visitors were well fed. By the beginning of the 1890s the ranch included a large house as well as storerooms, corrals, stables, and barns.

A post office opened at the ranch on July 15, 1891, to serve all of Pahrump Valley. In 1905 Harsha White purchased the ranch. Yount died soon afterwards, in 1907. He and Nehemiah Clark operated a sawmill in Clark Canyon, ten miles east on the northwest side of Mount Charleston. The sawmill, which supplied wood for the railroad ties that were being laid to Bullfrog and Rhyolite, continued to operate until 1915. Regular stages from Ivanpah to Bullfrog stopped at Manse Ranch for water and food. The population of Manse Ranch and the immediate area stood at fifty in 1911, and this figure remained fairly constant for quite a while. The post office closed on March 31, 1914, but the ranch continued to function. The stage lines stopped running in the teens, and the ranch became a solely agricultural enterprise. When Dr. H. D. Cornell bought the ranch in the 1930s, however, it had been abandoned. The ranch is still in use but has been incorporated into a dairy farm established more recently adjacent to the old Manse Ranch.

Meikeljohn

DIRECTIONS: From Beatty, head south on US 95 for 1 mile. Exit left and follow the road for another mile. Bear left and continue for 6½ miles. Exit left and follow this road for 2 miles to Meikeljohn.

Meikeljohn, a short-lived camp, formed during the same period as did nearby Telluride and Fluorine. The camp was named after George Meikeljohn, a popular politician of that time. It lasted only a few months, and nothing of value was ever found there. Meikeljohn never progressed beyond the tent-camp stage, and prospectors took everything with them when the camp died. Nothing at all remains.

Mexican Camp

DIRECTIONS: From Bullfrog, continue west for 6½ miles, ignoring all roads exiting to the right. At a fork, take a right and follow the road for 4½ miles to Mexican Camp.

Mexican Camp was a small, short-lived mining camp that sprang up during the early 1900s. The camp was a result of the frantic exploration following gold discoveries in nearby Rhyolite and Bullfrog. Almost overnight a small tent camp formed. Its peak population was about twenty-five. The gold faded quickly, and Mexican Camp was abandoned only three months after it had formed. Nothing remains of the site except some shallow test-mining holes.

Mud Spring Station

DIRECTIONS: From Beatty, head north on US 95 for 11 miles. Exit left and follow this road for 10 miles. Exit left and follow the road for 5 miles to Mud Spring Station.

Mud Spring Station was a water stop for the Las Vegas and Tonopah Railroad. The stop was on top of Mud Spring summit, where trains needed to replenish their water supply after climbing the steep grade coming out from Rhyolite and the Amargosa Flats. A pipeline from springs on the nearby hillside kept the tank full of cool water to soothe the strained boilers of overworked locomotives. A near tragedy took place in Mud Spring Station in March 1909 when J. A. Burke, the section boss, was almost knifed to death during a robbery attempt, but he survived. The stop was abandoned when the Las Vegas and Tonopah Railroad stopped running in 1914. Nothing marks the site except a small group of green trees and four foundations.

Original

(Orion) (Amargosa) (Bullfrog) (Aurum)

DIRECTIONS: *From Bullfrog, continue westward for 1½ miles. Bear left and continue for 1 mile. Bear right and continue for ¾ mile to Original.*

Original, first called Amargosa, was a small settlement located near the original Bullfrog mine. It was here that excitement over the Bullfrog district began to grow when Frank "Shorty" Harris and Ernest Cross made their discoveries on August 4, 1904. It is not known whether their discoveries were the first made in the area. There is evidence that George Ladd and Otis Johnson were the actual discoverers and that Harris and Cross came weeks later but took credit for the initial findings. A small camp, called Orion, or Aurum, formed at the mine. A separate tent camp, located on the flat below the mine, also formed and was named Amargosa. Bullfrog, located half a mile below Amargosa, sprang up as well. Later, when the Las Vegas and Tonopah Railroad established a stop just south of the mine, the Amargosa townsite was renamed Original. Shortly after the discoveries were made, Doc Benson laid out the Amargosa and Aurum townsites, and Milton Detch laid out the Bullfrog townsite (a camp separate from the town of the same name that was to form later near Rhyolite). Amargosa quickly developed into the major camp of the three. Its first store, run by Len McGarry, opened in October. By December Amargosa had five stores, seven saloons, a restaurant, a post office, and the Hotel Courtmarsh. All of the businesses were housed in tents. By February a phone line ran to Goldfield, and the Amargosa Trading Company formed. In March Prudden and Robbins began running a stage to Las Vegas, and the camp had a population of 100. Unfortunately, the Amargosa Townsite Company and the Bullfrog District Water and Ice Company hadn't found a convenient source of water. The cost of hauling barrels of water in was extremely high. The townsite company elected to purchase the new Bonanza townsite and renamed it Bullfrog. Basically, the entire tent camp was picked up and moved to the new Bullfrog. By April there was only one tent left in Amargosa. A dozen were still at the old Bullfrog camp, but by the middle of summer everyone had moved to Bullfrog and Rhyolite.

Once Bullfrog and Rhyolite began to boom, the tent town of Amargosa moved to Bullfrog. Both Harris and Cross sold their shares of the Bullfrog mine soon after the discovery was made. Cross sold his half for $25,000. Harris got drunk and sold his share for a mule and $500. The Original Bullfrog Mines Syndicate controlled the mine and continued to operate it until the Panic of 1907 forced it to close. The Reorganized Original Bullfrog Mines Syndicate's attempt to reopen the mine in 1917 never got off the ground. The property was sold at a tax sale in 1918 and remained out of operation until 1924, when a few leasers worked it.

The newly formed New Original Bullfrog Mines Company purchased the mine in 1926 and leased it to the Bullfrog Mines Company. When that company folded in November 1927 the New Original Company regained control of the mine. In January 1928 the Nevada Operating Company took a twenty-year lease on the mine, did some work, and built a four-mile water pipeline from Indian Springs. The company folded in 1929, and the New Original Bullfrog Company once again regained control.

The company finally left the district, and J. Burmeister and W. S. Ballinger, both from Auburn, California, began to work the mine in the early 1930s. They renamed it the Burmball mine and set to work retimbering the 220-foot shaft. By time they finished working the mine in the late 1930s, it was 250 feet deep and had 2,400 feet of lateral work.

The mine has been worked periodically since the 1930s, but no recorded production has taken place. Absolutely no ruins remain at the Orion, the Bullfrog, or the Amargosa sites, for they never progressed beyond the tent-camp stage. There are some scant remains at the mine, including the hoist and one dilapidated building. The site is very hard to reach and is not worth the effort necessary to get there.

Pioneer

(Mayflower) (Crystal Springs) (Fountain City) (Giles Camp) (Bryan Camp)

DIRECTIONS: *From Beatty, head north on US 95 for 5½ miles. Exit left and follow the road for 1¼ miles. Take the right fork and follow it for 2 miles. Exit left and follow the road for 1 mile to Pioneer.*

Activity in the Pioneer area began in early 1906, before the town of Pioneer started to take shape. The first camp in the area formed at Crystal Springs, about half a mile below the future Pioneer townsite. George Thatcher made the first claims in March. He sold the claims to W. F. Gann and Howard Hinkle for $75,000 in April. By May the camp had been given the name Fountain City. In 1906 its active mines included the Mayflower, the Starlight, the Ziegler, the Gold Frog, the Aurora Bullfrog, and the Wellington. The Mayflower was the richest of these, and another camp, called Mayflower, formed near the mine. The Bullfrog Mining and Water Company opened an assay office. George Thatcher platted a townsite and built a road to Rhyolite. The first business was a boardinghouse owned by James O'Connell. By the spring of 1907 the Croesus, the Bullfrog Gold Reef, the Valley View, and the Hinkle mines had also opened. The discovery of the Pioneer mine, however, changed the district forever. A new camp began to form in 1908 between the Pioneer and the Mayflower mines. It grew at an amazing rate, acquiring a popula-

tion of 1,000 by 1909. A post office, with Edward Harlebrath as postmaster, opened on March 2, and Pioneer was officially a town. Two newspapers, the *Pioneer Press* and the *Pioneer Topics,* started publication in early 1909 but folded in August 1909.

At Crystal Springs two separate camps still existed despite the nearby boom. Bryan Camp and Giles Camp survived into 1910. The Bryan Camp was named for J.R. Bryan, president of the Little Ruth Mining Company, which had a couple of nearby mines. The Giles Camp was named for E.S. Giles, who came from Cripple Creek and who was superintendent of the Puritan and Gold Reef Mining Company.

By February 1909 Pioneer had two sections called upper and lower towns. Businesses included the H.D. and L.D. Porter store; the Pioneer Lumber Company, owned by Kuhlman and Ellis; the Lyric Theater; the Knotzer Cigar Store; the Northern Saloon, owned by French and Brumbly; the 66 Club, owned by Taylor and Remick; the Star Hotel, owned by C.N. Wickoff; the Pioneer Club, owned by Marie West; the Star Bakery, owned by Van Gelder and Simburger; the Pioneer Restaurant, owned by Jacob Eck; the Vienna Bakery and Cafe, owned by Ferar and McCourt; the Hotel Cecil, owned by Mrs. Courtma; the Pioneer Boot Store, owned by Wilkerson and Hutchinson; the Broadway Hotel, owned by Ham Chambers; the Pioneer Miners Boardinghouse, owned by S.B. Oldham; the Western Union office; the forty-room

Miners Hotel, owned by Tex Rickard; the Holland House; the McCormick Drug Company; and the Pioneer Bank. The Pioneer Mining and Townsite Company, run by George Von Polena, planned a railroad spur to Springdale. Alkali Bill Brong ran an auto line to Springdale. The Las Vegas and Tonopah Railroad also had an auto line running to Rhyolite at a cost of two dollars and fifty cents. A school was built, and the Nevada-California Power Company brought in electricity. At this time the mines were reportedly producing $200,000 worth of ore a month, but that figure seems highly exaggerated. In August 1908 the Pioneer mill was completed, and in May 1909 the five-stamp Mayflower mill began production.

The Pioneer Consolidated Mines Company purchased the Pioneer mine and eleven other claims in 1909. William Tobin was president of the company, which was incorporated in May in Wyoming and which had mine offices in Pioneer. George Wingfield was a major stockholder. The Pioneer mine was 430 feet deep and had more than 15,000 feet of branch tunnels. The ore, assaying at up to $20 a ton, consisted mostly of gold and also had minor values in silver. In August 1913 the company completed a ten-stamp mill along with a thirty-ton cyanide and amalgamation mill to treat the Pioneer mine ore. The mine partially collapsed in December 1914 and was forced to shut down along with the mills.

A fire that swept through the tinder-dry wood buildings in 1909 was devastating for the booming town. Although the town was soon partly rebuilt, the fire took the life out of Pioneer, and it never really recovered. The fire, which started in Bill Brong's office when his wife knocked the pipe out of a stove, completely destroyed all the businesses in the upper part of town. Damage amounted to $60,000. The residents were very bitter because Brong had been warned about the unsafe stove before the accident happened. Another big blow to the town came in the form of the collapse of the Pioneer Bank in July 1909. Many local residents lost their savings when the bank folded. By 1911 the population had fallen to 300, and the downhill trend continued.

The Consolidated Mayflower Mine Company gained control of the Mayflower mine and ten other claims after the fire. The mine area contained two tunnels, both of which were over 1,000 feet long, and four shafts ranging in depth from 100 feet to 350 feet. The ore was not quite as rich as was that of the nearby Pioneer mine, assaying at an average of $17 per ton. The Mayflower mine closed down in 1928 after producing $400,000 worth of ore. The company also successfully reworked the Starlight mine. Early on, in December 1909, the company was forced to shut down due to A.C. Eisen's poor management. He came under suspicion when $11,000 in bullion came up missing. Eisen, in the end, formed the Mayflower Leasing Company and leased the Mayflower mine to his new company for twenty-five years. Needless to say, the stockholders were in an uproar. The ensuing litigation effectively put a stop to work on the Mayflower mine for years. The company wasn't able to resume operations until January 1919.

The reopening of the Pioneer mine in May 1915 gave sagging Pioneer a small lift. The Pioneer Consolidated Mines Company reorganized in May 1918 and was appropriately renamed the Reorganized Pioneer Mines Company. It continued working the Pioneer mine until 1931. The old ten-stamp Smith mill at Beatty moved to Pioneer in early 1915. During its period of operation, the mine produced more than $500,000 worth of gold and silver.

A smaller company, the Indiana Mines Exploration Company, was also active in Pioneer for a while. The company, which owned claims adjoining the Mayflower and Pioneer mines, reopened the Mayflower in 1916. It sank a 500-foot shaft and in May 1929 completed a 75-ton amalgamation mill. In 1930 the company merged with the Reorganized Pioneer Mines Company. William Tobin was still a major influence in the district. He was president of both the Pioneer Consolidated Company and the Consolidated Mayflower Company. When the Reorganized Pioneer Mines Company consolidated all the holdings, Tobin became its president, too.

After the Pioneer mine closed in 1931, the town died. The post office ceased to operate on January 26, 1931. The Indiana Company's mill moved to Gilbert in March. In December 1936 the General Milling Company, which was also active in Rhyolite, bought the Pioneer and the Mayflower mines and in March 1937 restarted the old Pioneer mill. The company's success was limited, however, and by 1939 the Pacific States Mining Corporation was running the mines, although they gave up the next year. After 1940 activity ceased at Pioneer. The district had produced a total of $1.6 million worth of ore. The ruins of Pioneer are scant. Only one building remains, along with the ruins of the mines and mills. The harsh desert climate has practically obliterated the site. To find the townsite layout, which is strewn with rubble and faint sunken foundations, requires a long search. At Crystal Springs there are a couple of buildings left. At the site of Giles Camp one building remains, and only foundations and ruins mark the Fountain City site.

Rhyolite

DIRECTIONS: *From Beatty, head west on Nevada 374 for 2 miles. Exit right and follow the road for 2.3 miles to Rhyolite.*

Although Rhyolite had a relatively short life, its story of dramatic rise and swift fall is one of the most fascinating of any ghost town in Nye County. Rhyolite formed soon after Frank "Shorty" Harris and Eddie Cross made rich discoveries in the summer of 1904 in the hills west of what would eventually be the townsite. Soon a small camp called Bullfrog sprang up.

Another camp, Rhyolite, formed a mile to the north. Rhyolite was staked out in November 1904 and officially platted on January 15, 1905. Within a month an unknown person, rumored to be Bob Montgomery, was offering merchants free lots. A small tent city materialized. It included numerous saloons, restaurants, and boardinghouses. One of the first substantial buildings constructed was the $30,000, two-story Southern Hotel.

An unofficial post office was established in early 1905 in Len McGarry's general store. In February it moved to Bill Porter's grocery store on growing Golden Street. He "delivered" the mail by yelling out the names of the addressees, an extremely inefficient process that normally took hours. The official U.S. post office, housed in a ten-by-twelve-foot tent, opened on May 19, 1905. Anna Moore was postmaster. With its increasingly large clientele, the post office outgrew a number of new offices before moving into the basement of the John S. Cook Bank building in 1908.

Water, a rare commodity in the Rhyolite area, was carted in at a cost of $2 to $5 per barrel. It was not until June 26, 1905, that Rhyolite had an efficient water system. The Indian Springs Water Company formed in 1905 and was soon piping water from Indian Springs, five miles to the north. Six weeks after the Indian Springs Company reached Rhyolite, the Bullfrog Water Company completed a water system pipeline to Rhyolite from Goss Springs, twelve miles away. This pipeline had a daily flow of 200,000 gallons. A short time later a third water company, the Bullfrog Water, Light and Power Company, also became active in the district. The company controlled thirteen springs whose total flow was one million gallons a day. Water pressure was strong enough to support a fire hydrant system with seventy pounds of pressure. Only a year after Rhyolite had been bone dry, the town had an abundance of the precious liquid. Two 20,000-gallon water reservoirs were built at the top of Golden Street.

In March 1905 a number of small camps other than Bullfrog and Rhyolite were developing within a radius of a few miles. Squattersville, a small tent city, was set up between Rhyolite and Bullfrog and eventually merged with the southern part of rapidly expanding Rhyolite. A mile from Bullfrog were Orion and Amargosa, tent camps with eighty tents and a population of 160.

Rhyolite was at its peak in January 1908. The major buildings shown are the Overbury Block and the John S. Cook Bank (both in center) and the solid concrete jail (lower right corner). (Nevada Historical Society)

Preserving the Glory Days

Bonanza, another small camp, was at the south end of Ladd Mountain. All of these faded quickly as attention focused on booming Rhyolite.

By the spring of 1905 four stage lines were bringing supplies to Rhyolite. The best known among these was the Kitchen stage, which brought supplies from Goldfield, eighty miles north, at a cost of $18 a ton. The first auto stage, run by the Tonopah and Goldfield Auto Company, became active in May 1905. Judge William Stewart, a famous lawyer, moved to the Rhyolite area that spring, bestowing a sense of prestige on the growing city. Baseball became the town's sports entertainment. The first game took place on June 11, 1905. Rhyolite beat Beatty ten to six. The town supported three teams, the 66 Club Tigers, the Rhyolite Stars, and the Railroad Bullets. Rhyolite built its first school in early 1906, and the enrollment soon reached 90. The school was blown off its foundation in September 1906 but was soon reset. By May 1907 the number of pupils had swelled to 250. Soon a $20,000, two-story brick school was built, with three classrooms on the first floor and one classroom and an auditorium on the second.

Rhyolite reached its peak in 1907 and 1908. Its population at that time is estimated to have been anywhere from 8,000 to 12,000. During this period two weekly newspapers and one daily competed for favor in Rhyolite. The weekly *Rhyolite Herald* was the first in town and was also the last to fold. It began publication on May 5, 1905, and continued until June 23, 1912. The *Bullfrog Miner,* after being published in Bullfrog for a little over a year, moved to Rhyolite on March 30, 1906. The *Rhyolite Daily Bulletin* began publication on September 23, 1907, but did not last long, folding on May 31, 1909. Two magazines were published in Rhyolite toward the end of the town's zenith.

The first was the *Death Valley Prospector*, which was first issued in November 1907 and was renamed the *Death Valley Magazine* the following month. It came out monthly until it folded in October 1908.

Rhyolite was served by three railroads during its peak years, an honor rarely bestowed on any Nevada city. The first railroad to reach Rhyolite was the Las Vegas and Tonopah, which began regular service on December 14, 1906. Then Senator William Clark and his relative, J. Ross Clark, strongly backed the railroad. John Brock of Tonopah ran the Bullfrog-Goldfield Railroad, the second railroad to arrive in Rhyolite. He had already made a name for himself with the prominent Tonopah and Goldfield Railroad. The Bullfrog-Goldfield was the weakest of the three railroads, but it was the first to complete a depot in Rhyolite. The depot was twenty-four feet by seventy-two feet and was finished in April 1908, just before the Las Vegas and Tonopah completed its own fancy depot. The third of the Rhyolite railroads, the Tonopah and Tidewater, never went as far as Rhyolite but had a station at nearby Gold Center. That station opened in October 1907, and the newly formed Tonopah and Tidewater soon became the best-established railroad in Nye County, lasting into the 1940s.

In January 1907 a network of 400 electric streetlight poles were installed, and soon Rhyolite was brightly lit twenty-four hours a day. By March 1907 Rhyolite's post office had the seventh-largest clientele in Nevada. A number of very impressive buildings were erected in the town in 1907 and early 1908. These included the $90,000 John S. Cook Bank building and the $50,000 Overbury building. H.D. and L.D. Porter, who operated mercantile stores throughout Nevada, built a new store in Rhyolite in 1907. The fancy stone building featured huge store windows, which were the talk of the town. At its peak Rhyolite had forty-five saloons, an opera house, a number of dance halls, two electric light plants, a two-story Miners' Union Hall with a sixty-foot front, a slaughterhouse, two railroad depots, numerous stores, countless other buildings—both wood and brick, and three public swimming pools. The Rhyolite Foundry and Machine Supply was organized in June 1907 to build cars. Parts were brought in, but the company folded before production began. Rhyolite even had its own stock exchange for a while in 1907.

During Rhyolite's brief reign of glory, more than eighty-five mining companies were active in the hills around the city. There were seven major mines: the Montgomery-Shoshone, the Denver, the National Bank, the Eclipse, the Polaris, the Gibraltar, and the Tramp. Of these, the Montgomery-Shoshone was the most productive. A Shoshone Indian that Bob Montgomery had sent to stake a claim for him discovered the mine. When Montgomery began to work the claim he kept running into talc deposits, which frustrated him. But one day, after a heavy rain, he happened to take a look at the dissolving talc pile and found gold. The talc ore assayed from $3,000 to $5,000 a ton. Soon afterward, Montgomery sold the mine to Charles Schwab for a reported $5

The statuesque ruins of the Cook Bank building. (Shawn Hall collection, Nevada State Museum)

million. Schwab then built a large mill, which started up in September 1907, and also convinced the Las Vegas and Tonopah Railroad to run a rail line by the mine, making it worthwhile for Schwab to mine lower-value ore because he incurred no extra cost for transportation. In October 1907 alone the mill produced $175,000 worth of ore.

The financial panic of 1907 killed Rhyolite. Most of the town's investors were from the East, and when they withdrew their backing all the mines were forced to close. The devastating effects of the panic did not affect Rhyolite until the spring of 1908. Then the trains were almost always filled to capacity with people leaving town. The city emptied as fast as it had filled only a few years earlier. On August 19, 1908, Rhyolite suffered a fire that leveled the red-light district and spread to parts of the eastern business district. By the end of 1909 the population was well below 1,000. Despite this, the residents pushed for the formation of Bullfrog County in the legislature, but the town's quick decline dashed these hopes. The Montgomery-Shoshone mine, Rhyolite's last real hope, closed after producing close to $2 million worth of ore. After a short closure, the mine and mill reopened in April 1910 but closed for good in March 1911. During this period seventy-five men were employed. The mine produced $246,000 worth of ore but at a cost of $244,000. The only mine left in operation was the old Tramp mine, just north of Bullfrog. The Sunset Mining and Development Company worked the mine for a while but finally gave up in the early teens. The railroads continued to operate, and the clientele continued to decline until an average of only two or three people took the trains each day.

The population had shrunk to 675 by 1910 and continued to fall rapidly during the next two years. Streetlights were shut off on April 30, 1910, although the power company operated until 1916. Earle Clemens, editor of the *Rhyolite Herald,* wrote a touching editorial published in the April 8, 1911, issue, his last before leaving to work in California:

> GOODBYE: It is with deep regret that I announce my retirement from the newspaper field in the Bullfrog district. It has been my lot to remain here while all my erstwhile contemporaries have fled, one by one, to more inviting localities, and now it is my time to say goodbye. May prosperity follow you everywhere, and catch up with you, too, and may prosperity again reign in Rhyolite—the prettiest, coziest mining town on the great American desert, a town blessed with ambitious, hopeful, courageous people, and with a climate second to none on earth. Goodbye, dear old Rhyolite.

After Clemens left, the Rhyolite Printing Company continued to publish the paper until it finally folded on June 22, 1912. All subscriptions were transferred to the *Goldfield News.*

The Las Vegas and Tonopah Railroad and the Bullfrog-Goldfield Railroad consolidated in July 1914 in an effort to maintain service to Rhyolite, but it was a losing cause. The railroad struggled on for a few more years but finally ended service in 1918. The last freight train left Rhyolite on August 17, and the last work train left on October 31. The rails were torn up the next year. The post office continued to operate until the population fell below twenty-five, when the postmaster, H. D. Porter, decided that it was no longer needed. While his store had closed in 1910, he had stayed on, hoping for a new boom. The office closed on September 15, 1919. The population of the almost-dead town had shrunk to fourteen by the beginning of 1920. The last resident, J. D. Lorraine, died in 1924.

During the 1920s and 1930s Rhyolite remained virtually unchanged. The only semblance of a revival took place in 1928, when the Rhyolite Consolidated Mines Company formed and began work on a number of mines on Bonanza Mountain. Capitalists from Virginia and Georgia backed the company, but they withdrew their support in 1930, and the company folded. Other attempts made during the next twenty years mainly consisted of re-working old tailing piles. Total production value for the Rhyolite mines, excluding current operations, is $3 million. The empty buildings in Rhyolite were full of furniture, and the bank floors were covered with official records and worthless stock. The only building occupied during this period was the old Las Vegas and Tonopah depot. Wes Westmoreland had purchased it in 1925 and had opened a casino and bar, called the Rhyolite Ghost Casino, in October 1937. The other buildings in town were left to the mercy of the harsh desert wind.

Westmoreland's sister, Mrs. H. H. Heisler, maintained the railroad station until recently. She ran a small museum and curio shop in the station. The bottle house, constructed of 51,000 beer bottles, is one of the few other buildings still in fairly good condition. However, it has been damaged during the past few years, either by the nearby mining operations or by a strong dust devil, depending whom you talk to. In addition to a few small wood structures, the only other substantial ruin is the jail. The more impressive remains in Rhyolite include the Cook Bank, the $20,000 school, and the Porter store. The front of the Cook building was altered to a Spanish style by a movie company. Grills drilled into the window frames seem to have weakened the front, and a large section collapsed the following year. The Rhyolite-Bullfrog cemetery is half a mile south of the Bullfrog townsite. It contains many interesting gravestone styles, including numerous wooden headboards. In 1986 the Stonewall Park Development Corporation attempted to purchase the ghost town to form the nation's first homosexual community. The county soundly rejected the idea.

A new organization, the Friends of Rhyolite, is now caring for the town. The organization has done an admirable job of helping to preserve what is left of the town and of protecting it from further vandalism. An annual celebration of Rhyolite is held in June and helps to raise funds for further work. Recently mining operations have begun near Rhyolite. Bond International Gold poured the first gold bar from its open pit operation on July 25, 1989. It produced 200,000 ounces of gold during the first year. The operation has continued to grow since then, and the huge mine dumps are located below the town. Barrick is now running the mine. From 1989 through 1994 it

The Rhyolite bottle house in better days. The Friends of Rhyolite are currently trying to restore this unique vestige of the past. (Southerland Studios, Carson City)

produced 1.44 million ounces of gold and 1.58 million ounces of silver. The company employs 290 workers.

It is almost unimaginable that the desolate site that used to be Rhyolite was once filled with more than 10,000 people and row upon row of buildings. Onlookers will feel a sense of emptiness, amazement, and even shock at the total devastation that has almost completely leveled the once-bustling city. Rhyolite is clearly one of the best ghost towns in Nye County and in the state, not just because of the buildings that remain but also because of the scope of change that has taken place there. If there is one place that exemplifies how fast life can change in a mining town, Rhyolite is that place. This was one of the first sites the author visited on his initial trip to Nye County in 1979, and it deserves credit for fostering his interest in ghost towns. Don't miss Rhyolite.

Rose's Well
(Palmer's Well)

DIRECTIONS: *From Beatty, head south on US 95 for 15½ miles. Exit right and follow this faint road for ¼ mile. Exit left and follow the road for ½ mile. Take a left and follow this road for ¾ mile to Rose's Well.*

G.W. Rose and E.E. Palmer, both of whom ran a small freight line from Las Vegas to Beatty, dug Rose's Well in the early 1900s. Their 210-foot well yielded more than 100 barrels of cool water each day. The well was known as both Rose's Well and Palmer's Well until Palmer's death. At the stage station, not only water but good meals were available, as was shelter for horses and tired travelers. Rose's Well was the only spot between Beatty and Las Vegas that offered such commodities.

When the Las Vegas and Tonopah Railroad was built, Rose's Well became an important way station. The railroad reached Rose's Well in July 1906. Before further construction could be undertaken, a tie shippers' strike stopped all work. Rose's Well was the end of the line for almost all of July. This created a flow of travelers going to and from Rhyolite and led to further development at Rose's Well.

The only killing at Rose's Well took place in July, when Bob Selby shot and killed a drunk who was threatening a woman. The killing was ruled justifiable homicide. After railroad construction resumed, the small settlement continued to be an important stop until the railroad reached Gold Center. The railroad constructed a number of buildings at Rose's Well in 1907 during the early days of the Lee boom. Then the station was downgraded to a water stop. On its timetable the railroad listed the stop as Roswell. The station and small complex continued to function until the Las Vegas and Tonopah folded and tore up its rails in 1918 and 1919. Rose's Well was then quickly abandoned.

Today Rose's Well is virtually nonexistent. The only marker is the old well, which is now dry. The site is so extremely dry and desolate that it is hard to imagine a busy railroad station once existed here.

South Bullfrog

DIRECTIONS: *South Bullfrog is located 1 mile northeast of Gold Center at Amargosa Narrows.*

South Bullfrog can't be called a town. It was rather a promoter's dream. B.V. Laughlin, who also planned to build a mill to treat ore from mines on Bare Mountain, platted a townsite at South Bullfrog in January 1906. The first business in South Bullfrog was Sam Johnson's saloon. A couple of other tents went up in February, but when Laughlin couldn't secure financing for his mill, dreams of South Bullfrog developing into a town evaporated, and everything was gone from the townsite by March. Nothing permanent was ever built, so nothing is left today.

Springdale

DIRECTIONS: *From Beatty, head north on US 95 for 9½ miles to Springdale, located on the east side of the highway.*

Springdale was a train stop on the Bullfrog-Goldfield Railroad, but the meadowy area had been occupied long before the steel ribbons were laid. During the nineteenth century Indians camped at the site. After they left, a number of small ranches were built in the fertile spot. The railroad's arrival in the winter of 1906 put Springdale on the map. A station house and a water tank were built, for the site had a very good water supply, and the steam engines of that era needed water. A. L. Lidwell platted a townsite on the Indian Joe Ranch in July.

A small settlement grew up around the railroad station, and a post office, with Albert Lidwell as postmaster, opened on February 19, 1907. Springdale received another big boost that year when a boom at nearby Pioneer brought a small rush of prospectors to the area. Springdale became the railhead for booming Pioneer and was soon enjoying a brisk, prosperous shipping business. By the summer of 1907 Springdale had four saloons, a number of restaurants, a hotel, a livery stable, and a popular red-light district housing fourteen women. In 1907 Springdale held an extremely large Independence Day celebration that included a Wild West show from Goldfield. More than

1,000 people from surrounding mining camps showed up, and everyone had a good time. The Springdale Mining and Milling Company, of which T.C. Rankin was president, started up a fifty-ton mill in February 1909. It treated ore from the nearby Newton-Wolfe, Aurora-Bullfrog, Forbes and Nation, and Oasis Bullfrog mines. In April 1909 Effie Clinkscales died. It was the town's first death, and it created the need for a cemetery. By May a school opened at Springdale in response to demand.

Springdale remained an important stop on the Bullfrog-Goldfield Railroad until Pioneer began to decline late in 1911. At the beginning of that year, Springdale still had a population of seventy-five. However, the subsequent closure of the mill was a tough blow. By the end of 1911 Springdale's population had shrunk to around ten. The post office closed on January 15, 1912, and soon only trains disturbed the town's slumber. Even the trains stopped running in 1928, and Springdale faded into history. A.L. Lidwell completed a cyanide plant in June 1938 to rework the tailings from the Pioneer and Mayflower mines at Pioneer, but in August, before operations could get fully underway, Lidwell died, and the plant closed.

A ranch has been active for many years at Springdale, and a few gas stations served travelers until 1958, when the new US 95 bypassed the area. The remaining buildings are on the ranch.

Stirling

(Sterling) (Timber Mountain)

DIRECTIONS: *From Amargosa Valley, head east on US 95 for 30 miles. Exit right and follow the road for 7 miles to Stirling.*

Stirling was an extremely short-lived mining camp. It lasted only from 1908 to 1909. There had been activity in the mountains around Stirling as early as 1869. After a short while, the Timber Mountain Mining District formed. It encompassed the north end of the Spring Mountain range. The Sterling mine was opened, and a small five-stamp mill was constructed. But activity was extremely limited, and the district was abandoned in 1870. Some additional activity took place in the 1890s. The Sterling Mining Company built an arrastra in 1890. Production continued until 1896. In November the company's owner and superintendent, a man named Gillespie, was shot and killed while walking between the Johnnie and the Chispas mines at Johnnie. No one ever found a suspect or determined a motive for the slaying.

The town of Stirling formed in April 1907 when the Ore City Mining Company reopened the Sterling mine and began intensive operations. The company found both gold and copper ore in the old mine and in other newly discovered claims in the area. A townsite was platted at the base of Mount Sterling in April 1907. There were soon twenty-five people living in the town. Supplies were brought in from the Las Vegas and Tonopah Railroad, seven miles to the north. Water came from Big Timber Spring, on the back side of Mount Sterling. In its May 3, 1907, issue, the *Reno Evening Gazette* remarked that Stirling appeared to be a promising camp. A post office, with Charles Knotts as postmaster, opened on June 3, 1907, just as the small boom at Stirling was dying out. The town struggled into 1908. The post office, however, closed on February 13, and Stirling ceased to exist shortly afterwards.

Only scant ruins are left at Stirling. No buildings remain, but rubble is scattered over the site. There are some meager mine ruins on the side of Mount Sterling. The road to Stirling is extremely rough and is impassable during wet weather.

Telluride

DIRECTIONS: From Beatty, head south on US 95 for 1 mile. Exit left and follow the road for 1 mile. Bear left and continue for 2¾ miles. Exit right and follow the road for ¼ mile. Exit left and follow this road for 2 miles. Exit right and follow this road for 1½ miles to Telluride.

Telluride was one of the many small camps that sprang up in the hills east of Beatty during the early twentieth century. Telluride was founded around the same time as was nearby Fluorine, and for a while the two camps were rivals. Initial discoveries were made in April 1908, and shortly afterwards John Doser platted a townsite. The camp was named Telluride because of the colorful tellurium found in the ore. In October J.P. Randall, D. Wiggens, and A.S. Pritchard sold fifteen claims for $50,000, but the new owners found nothing. Telluride slowly disappeared. The camp last appeared on maps in 1910, although miners worked some of the mines until 1915. In September 1931 George Wingfield bought Emily Kieron's cinnabar claims, which had been discovered in 1909, for $25,000. Five men were employed, but Wingfield gave up in 1932. During its peak, Telluride had a population of twenty-five. Nothing substantial was ever built there, and only some small tailing piles mark the site. Some recent activity has brought about the only real production Telluride has ever seen. In 1989 and 1990 the open-pit Motherlode mine produced 30,000 ounces of gold.

Transvaal
(Nyopolis) (Gold Gulch)

DIRECTIONS: From Beatty, head north on US 95 for 4½ miles. Exit right and follow this poor road for 11 miles to Transvaal.

Transvaal was an extremely short-lived mining town that experienced one of the most incredible booms and one of the fastest declines of any town in Nye County history. E.S. Chafey and George Probasco, who opened the Transvaal and Eastern Contact mines, made initial discoveries in the northern section of the Beatty Wash in February 1906. A small tent camp, called Nyopolis, soon formed. By the beginning of April the tent camp had been renamed Transvaal, which meant "across the river" and was a clear reference to Transvaal's location across the Amargosa River from Beatty and Rhyolite. The Transvaal Mining District formed on April 4, and Joe Ray was named recorder.

By the second week of April there were more than seventy-five tents in Transvaal. The Transvaal-Nevada Mining Company controlled all the mines. R.W. Gorrill was the company's president, and Chafey was its manager. Business establishments at the camp included four saloons, an assay office, a lumberyard, a lodging house, and a number of broker and real estate offices. The Transvaal Townsite Company offered lots for $25 to $350. Two newspapers, the *Transvaal Miner* and the *Transvaal Tribune,* began publication in early April. Transvaal reached its peak during the third week of April, when 700 to 800 people were living in the tent city. The Transvaal Stage Company, owned by J.R. Knoll, ran two daily stages for Beatty and left from Transvaal every morning, charging $5 a person. Borax Smith began running an auto stage, and the Tonopah and Tidewater Railroad planned a six-mile spur to the camp.

The swift and dramatic collapse of Transvaal occurred in the first two weeks of May 1906. When it was discovered that little valuable ore existed at Transvaal, the people left in droves. Litigation then put an end to any remaining activity. Both papers folded in early May, and by the third week of that month the site was totally abandoned. Only Probasco remained. He sold the Transvaal Company's property to Al Meyers of Goldfield. The Desert Queen produced a little more than $17,000 worth of ore before Meyers gave up in 1910. Absolutely nothing remains in Transvaal except some mine dumps. The town did not last long enough for any permanent structures to be erected. The site is difficult to reach and is very unrewarding.

References

Northwestern Nye County

Advertiser (Ione). 1864.

Angel, Myron. *History of Nevada.* Oakland: Thompson and West, 1881. Reprint, Berkeley: Howell-North Books, 1958.

Armstrong, Robert D. *A Preliminary Union Catalogue of Nevada Manuscripts.* Reno: University of Nevada Library and Nevada Library Association, 1967.

Ashbaugh, Don. *Nevada's Turbulent Yesterday, A Study in Ghost Towns.* Los Angeles: Westernlore Press, 1963.

Bailey, Edgar, and David A. Phoenix. *Quicksilver Deposits in Nevada.* University of Nevada Bulletin vol. 38, no.5. Reno: Nevada State Bureau of Mines and MacKay School of Mines, 1944.

Ball, S. H. *A Geological Reconnaissance in Southwestern Nevada and Eastern California.* U.S. Geological Survey Bulletin no. 308. Washington, D.C.: Government Printing Office, 1907.

Basso, Dave. *Nevada Lost Mines and Hidden Treasures.* Sparks, Nev.: Dave's Printing and Publishing, 1974.

Bastin, Edson, and Francis Laney. *The Genesis of the Ores at Tonopah, Nevada.* Professional Paper no. 104. Washington, D.C.: Government Printing Office, 1918.

Beebe, Lucius, and Charles Clegg. *U.S. West: The Saga of Wells Fargo.* New York: E. P. Dutton, 1949.

Belmont Courier. 14 February 1874–2 March 1901.

Billeb, Emil. *Mining Camp Days: Bodie, Aurora, Bridgeport, Hawthorne, Tonopah, Landy, Masonic, Benton, Thorne, Mono Mills, Mammoth, Sodaville, Goldfield.* Berkeley: Howell-North Books, 1968.

Blatchly, A. *Mining and Milling in the Reese River Region of Central and Southeastern Nevada.* New York: Slate & Janes, 1867.

Bowers, Martha, and Hans Muessig. *History of Central Nevada.* Cultural Resource Series no. 4. Reno: Nevada State Office Bureau of Land Management, 1982.

Browman, Mickey. *Nevada Ghost Town Trails.* N.p.: Abbott & Abbott, 1973.

Brown, Mrs. Hugh. *Lady in Boomtown: Miners and Manners on the Nevada Frontier.* Palo Alto: American West Publishing Company, 1968. Reprint, Reno: University of Nevada Press, 1991.

Bruner, Firmin. *Some Remembers — Some Forgot: Life in Central Nevada Mining Camps.* Carson City, Nev.: Nevada State Park History Association, 1974.

Burt, Olive. *Jedediah Smith: Fur Trapper of the Old West.* New York: Julian Messner, 1951.

Carlson, Helen S. *Nevada Place Names: A Geographical Dictionary.* Reno: University of Nevada Press, 1974.

Carpenter, Jay A., Russell R. Elliott, and Byrd F. Sawyer. *The History of Fifty Years of Mining at Tonopah, 1900–1950.* University of Nevada Bulletin vol. 47, no. 1. Geology and Mining Series no. 51. Reno: Nevada State Bureau of Mines and MacKay School of Mines, 1953.

Central Nevada's Glorious Past. 1978–1996.

Clark, Lewis, and Virginia Clark. *High Mountains & Deep Valleys: The Gold Bonanza Days.* San Luis Obispo, Calif.: Western Trails Publishing, 1978.

Cook, Fred S. *Legends of Nye County.* Hawthorne, Nev.: Times, 1978.

Couch, Bertrand F., and Jay A. Carpenter. *Nevada's Metal and Mineral Production.* University of Nevada Bulletin vol. 37, no. 4. Geology and Mining Series no. 38. Reno: Nevada State Bureau of Mines and MacKay School of Mines, 1943.

Cronkhite, Daniel. *Ghost Towns of Esmeralda and Nye Counties, Nevada.* Van Nuys: High School Press, 1960.

Davis, Sam P., ed. *The History of Nevada.* Reno: Elms Publishing Company, 1913.

Douglass, William A, and Robert Nylen. *Letters from the Nevada Frontier, Correspondence of Tasker L. Oddie, 1898–1902.* Norman, Okla.: University of Oklahoma Press, 1992.

Elliott, Russell R. *Nevada's Twentieth-Century Mining Boom: Tonopah, Goldfield, Ely.* Reno: University of Nevada Press, 1966.

Emmons, William H. *A Reconnaissance of Some Mining Camps in Elko, Lander and Eureka Counties, Nevada.* U.S. Geological Survey Bulletin no. 408. Washington, D.C.: Government Printing Office, 1910.

Florin, Lambert. *Ghost Towns of the West.* New York: Promontory Press, 1971.

Folkes, John. *Nevada's Newspapers: A Bibliography, a Compilation of Nevada History, 1854–1964.* Reno: University of Nevada Press, 1964.

Fox, Theron. *Nevada Treasure Hunters Ghost Town Guide: Handy Reference to Locating Old Mining Camps, Ghost Town Sites, Mountains, Rivers, Lakes, Camel Trails, Abandoned Roads, Springs, and Water Holes.* San Jose, Calif.: T. Fox, 1961.

Frickstad, Walter N., and Edward W. Thrall. *A Century of Nevada Post Offices, 1852–1957.* Oakland, Calif.: Philatelic Research Society, 1958.

Grantsville Bonanza. 1881.

Grantsville Sun. 19 October 1878–16 April 1879.

Grover, David H. *Diamondfield Jack: A Study in Frontier Justice.* Reno: University of Nevada Press, 1968.

Hall, Shawn. *Romancing Nevada's Past, Ghost Towns and Historic Sites of Eureka, Lander, and White Pine Counties, Nevada.* Reno: University of Nevada Press, 1994.

————. *A Guide to the Ghost Towns and Mining Camps of Nye County, Nevada.* New York: Dodd Mead, 1981.

Harris, Robert. *Nevada Postal History.* Santa Cruz, Calif.: Bonanza Press, 1973.

Higgins, L. James. *A Guide to the Manuscript Collections at the Nevada Historical Society.* Reno: Nevada Historical Society, 1975.

Hill, James. *Mining Districts of the Western United States.* Bulletin no. 507. Washington, D.C.: Government Printing Office, 1912.

————. *Mining Districts—California and Nevada.* Bulletin no. 594. Washington, D.C.: Government Printing Office, 1915.

Kelly, J. Wells. *First Directory of Nevada Territory (1862).* Los Gatos, Calif.: Talisman Press, 1962.

Kleinhampl, Frank J., and Joseph I. Ziory. *Mineral Resources of Northern Nye County.* Nevada Bureau of Mines Bulletin 99B. Reno: Nevada Bureau of Mines and Geology, University of Nevada, Reno, 1984.

Kral, Victor. *Mineral Resources of Nye County, Nevada.* University of Nevada Bulletin vol. 45, no. 3. Geology and Mining Series no. 50. Reno: Nevada State Bureau of Mines and MacKay School of Mines, 1951.

Labbe, Charles. *Rocky Trails of the Past.* Las Vegas: privately published, 1960.

Lawrence, Edmund. *Antimony Deposits of Nevada.* Nevada Bureau of Mines Bulletin no. 61. Reno, Mackay School of Mines, 1963.

Lawson, William, and Paul Wesley. *Mines and Mining in Tonopah and Goldfield.* Tonopah, Nev.: privately published, n.d.

Lewis, Marvin. *Martha and the Doctor.* Reno: University of Nevada Press, 1977.

Lincoln, Francis Church. *Mining Districts and Mineral Resources of Nevada.* Reno: Nevada Newsletter Publishing Company, 1923.

Lingenfelter, Richard E. *The Newspapers of Nevada: A History and Bibliography, 1854–1979.* San Francisco: John Howell Books, 1964. Reprint, Reno: University of Nevada Press, 1984.

Mack, Effie. *Nevada: A History of the State from the Earliest Times through the Civil War.* Glendale: Arthur Clark, 1936.

McCracken, Robert. *A History Tonopah, Nevada.* Tonopah, Nev.: Nye County Press, 1990.

————. *Bibliography of the Histories of Nye and Clark County.* Tonopah, Nev.: Central Nevada Historical Society, 1991.

Moody, Eric. *An Index to the Publications of the Nevada Historical Society, 1907–1971.* Reno: Nevada Historical Society, 1977.

Morrissey, Frank R. *Turquoise Deposits of Nevada.* Nevada Bureau of Mines Report no. 17. Reno: Nevada Bureau of Mines, 1968.

Murbarger, Nell. *Ghosts of the Glory Trail.* Palm Desert, Calif.: Desert Magazine Press, 1956.

Myrick, David. *Railroads of Nevada and Eastern California.* 2 vols. Berkeley: Howell-North Books, 1963. Reprint, Reno: University of Nevada Press, 1992.

Nye County Historic Property Survey. Tonopah, Nev.: Nye County, 1980.

Nye County News (Ione). 25 June 1864–May 1867.

Paher, Stanley. *Nevada Ghost Towns and Mining Camps.* Las Vegas: Nevada Publications, 1970.

———. *Tonopah: Silver Camp of Nevada.* Las Vegas: Nevada Publications, 1978.

Quartz Mountain Miner. 16 June 1926.

Spurr, Josiah. *Geology of Nevada South of the Fortieth Parallel.* Bulletin no. 208. Washington, D.C.: Government Printing Office, 1903.

Stevens, Horace. *The Copper Handbook: A Manual of the Copper Industry of the United States and Foreign Countries.* Vols. 2–10. Washington, D.C.: Government Printing Office, 1908.

Stuart, E. E. *Nevada's Mineral Resources.* Carson City, Nev.: State Printing Office, 1909.

Tonopah Miner. 20 June 1902–5 November 1921.

Tonopah Times. 1 December 1915–15 November 1929.

Tonopah Times-Bonanza. 16 November 1929–31 December 1996.

United States Treasury Dept. *Report of J. Ross Browne on the Mineral Resources of the States and Territories West of the Rocky Mountains.* Washington, D.C.: Government Printing Office, 1867.

Vanderburg, William O. *Placer Mining in Nevada.* University of Nevada Bulletin vol. 30, no. 4. Geology and Mining Series no. 27. Reno: Nevada State Bureau of Mines and MacKay School of Mines, 1936.

Weed, Walter. *The Copper Handbook: A Manual of the Copper Mining Industry of the World.* Vol. 2. Washington, D.C.: Government Printing Office, 1912–1913.

———. *The Mines Handbook and Copper Handbook.* Vols. 12–17. Washington, D.C.: Government Printing Office, 1916–1926.

Northeast Nye County

Angel, Myron. *History of Nevada.* Oakland: Thompson and West, 1881. Reprint, Berkeley: Howell North Books, 1958.

Armstrong, Robert. *A Preliminary Union Catalogue of Nevada Manuscripts.* Reno: University of Nevada Library and Nevada Library Association, 1967.

Ashbaugh, Don. *Nevada's Turbulent Yesterday: A Study in Ghost Towns.* Los Angeles: Westernlore Press, 1963.

Bailey, Edgar, and David Phoenix. *Quicksilver Deposits in Nevada.* University of Nevada Bulletin no. 41. Reno: Nevada State Bureau of Mines and MacKay School of Mines, 1944.

Ball, Sidney H. *A Geological Reconnaissance in Southwestern Nevada and Eastern California.* Bulletin no. 308. Washington, D.C.: Government Printing Office, 1907.

Basso, Dave. *Nevada Lost Mines and Hidden Treasures.* Sparks, Nev.: Dave's Printing, 1972.

Beebe, Lucius, and Charles Clegg. *U.S. West, the Saga of Wells Fargo.* New York: E. P. Dutton, 1949.

Belmont Courier. 14 February 1874–2 March 1901.

Billeb, Emil. *Mining Camp Days.* Berkeley: Howell-North Books, 1968.

Blatchly, A. *Mining and Milling in the Reese River Region of Central and Southeastern Nevada.* New York: Slate & Jones, 1867.

Bowers, Martha H., and Hans Muessig. *History of Central Nevada.* Cultural Resource

Series no. 4. Tonopah, Nev.: Nevada State Office Bureau of Land Management, 1982.

Browman, Mickey. *Nevada Ghost Town Trails*. N.p.: Abbott & Abbott, 1973.

Brown, Mrs. Hugh. *Lady in Boomtown: Miners and Manners on the Nevada Frontier.* Palo Alto: American West Publishing Company, 1968. Reprint, Reno: University of Nevada Press, 1991.

Burt, Olive. *Jedediah Smith: Fur Trapper of the Old West*. New York: Julian Messner, 1951.

Carlson, Helen S. *Nevada Place Names: A Geographical Dictionary*. Reno: University of Nevada Press, 1974.

Central Nevada's Glorious Past. 1978–1996.

Clark, Lewis, and Virginia Clark. *High Mountains & Deep Valleys: The Gold Bonanza Days*. San Luis Obispo, Calif.: Western Trails Publishing, 1978.

Cook, Fred. *Legends of Nye County*. Hawthorne, Nev.: Times, 1978.

Couch, Bertrand F., and Jay A. Carpenter. *Nevada's Metal and Mineral Production*. University of Nevada Bulletin vol. 37, no. 4. Geology and Mining Series no. 38. Reno: Nevada State Bureau of Mines and MacKay School of Mines, 1943.

Cronkhite, Daniel. *Ghost Towns of Esmeralda and Nye Counties, Nevada*. Van Nuys, Calif.: High School Press, 1960.

Davis, Sam P., ed. *The History of Nevada*. Reno: Elms Publishing Company, 1965.

Emmons, William H. *A Reconnaissance of Some Mining Camps in Elko, Nevada*. U.S. Geological Survey Bulletin no. 408. Washington, D.C.: Government Printing Office, 1910.

Florin, Lambert. *Ghost Towns of the West*. New York: Promontory Press, 1971.

Folkes, John. *Nevada's Newspapers: A Bibliography, a Compilation of Nevada History, 1854–1964*. Reno: University of Nevada Press, 1964.

Fox, Theron. *Nevada Treasure Hunters Ghost Town Guide: Handy Reference to Locating Old Mining Camps, Ghost Town Sites, Mountains, Rivers, Lakes, Camel Trails, Abandoned Roads, Springs, and Water Holes*. San Jose, Calif.: T. Fox, 1961.

Frickstad, Walter N., and Edward W. Thrall. *A Century of Nevada Post Offices, 1852–1957*. Oakland: Philatelic Research Society, 1958.

Hall, Shawn. *Romancing Nevada's Past, Ghost Towns and Historic Sites of Eureka, Lander, and White Pine Counties, Nevada*. Reno: University of Nevada Press, 1994.

———. *A Guide to the Ghost Towns and Mining Camps of Nye County, Nevada*. New York: Dodd Mead, 1981.

Harris, Robert. *Nevada Postal History: 1861 to 1972*. Santa Cruz, Calif.: Bonanza Press, 1973.

Higgins, L. James. *A Guide to the Manuscript Collections at the Nevada Historical Society*. Reno: Nevada Historical Society, 1975.

Hill, James M. *Mining Districts of the Western United States*. Bulletin no. 507. Washington, D.C.: Government Printing Office, 1912.

———. *Mining Districts—California and Nevada*. Bulletin no. 594. Washington, D.C.: Government Printing Office, 1915.

———. *Notes on some Mining Districts in Eastern Nevada*. U.S. Geological Survey Bulletin no. 648. Washington, D.C.: Government Printing Office, 1916.

Kelly, J. Wells. *First Directory of Nevada Territory (1862)*. Los Gatos, Calif.: Talisman Press, 1962.

Kleinhampl, Frank J., and Joseph I. Ziory. *Mineral Resources of Northern Nye County*. Nevada Bureau of Mines Bulletin 99B. Reno: Nevada Bureau of Mines and Geology, 1984.

Kral, Victor. *Mineral Resources of Nye County, Nevada*. University of Nevada Bulletin vol. 45, no. 3. Geology and Mining Series no. 50. Reno: Nevada State Bureau of Mines and MacKay School of Mines, 1951.

Labbe, Charles. *Rocky Trails of the Past*. Las Vegas: privately published, 1960.

Lawrence, Edmund. *Antimony Deposits of Nevada*. Nevada Bureau of Mines Bulletin no. 61. Reno: Mackay School of Mines, 1963.

Lewis, Marvin. *Martha and the Doctor*. Reno: University of Nevada Press, 1977.

Lincoln, Francis Church. *Mining Districts and Mineral Resources of Nevada*. Reno: Nevada Newsletter Publishing Company, 1923.

Lingenfelter, Richard E., and Karen Rix Gash. *The Newspapers of Nevada: A History and Bibliography, 1854–1979*. San Francisco: John Howell Books, 1964. Reprint, Reno: University of Nevada Press, 1984.

Mack, Effie. *Nevada: A History of the State from the Earliest Times through the Civil War*. Glendale, Calif.: Arthur H. Clark, 1936.

McCracken, Robert D. *Bibliography of the Histories of Nye and Clark County*. Tonopah, Nev.: Central Nevada Historical Society, 1991.

McCracken, Robert D., and Jeanne Sharp Howerton. *A History of Railroad Valley, Nevada*. Tonopah, Nev.: Central Nevada Historical Society, 1996.

Moody, Eric. *An Index to the Publications of the Nevada Historical Society, 1907–1971*. Reno: Nevada Historical Society, 1977.

Morrissey, Frank R. *Turquoise Deposits of Nevada*. Nevada Bureau of Mines Report no. 17. Reno: Nevada Bureau of Mines, 1968.

Mountain Champion (Belmont). 3 June 1868–24 April 1869.

Murbarger, Nell. *Ghosts of the Glory Trail*. Palm Desert, Calif.: Desert Magazine Press, 1956.

Myrick, David. *Railroads of Nevada and Eastern California*. Vol. 2. Berkeley: Howell-North Books, 1963. Reprint, Reno: University of Nevada Press, 1992.

Nye County Historic Property Survey. Tonopah, Nev.: Nye County, 1980.

Nye County News (Ione). 25 June 1864–May 1867.

Paher, Stanley. *Nevada Ghost Towns and Mining Camps*. Las Vegas: Publications, 1970.

Silver Bend Reporter (Belmont). 30 March 1867–28 July 1868.

Spurr, Josiah. *Geology of Nevada South of the Fortieth Parallel*. Bulletin no. 208. Washington, D.C.: Government Printing Office, 1903.

Stevens, Horace. *The Copper Handbook: A Manual of the Copper Industry of the United States and Foreign Countries*. Vols. 2–10. Washington, D.C.: Government Printing Office, 1908.

Stuart, E. E. *Nevada's Mineral Resources*. Carson City, Nev.: State Printing Office, 1909.

Tonopah Miner. 20 June 1902–5 November 1921.

Tonopah Times-Bonanza. 16 November 1929–31 December 1996.

Tybo Sun. 19 May 1877–8 March 1880.

United States Treasury Dept. *Report of J. Ross Browne on the Mineral Resources of the States and Territories West of the Rocky Mountains.* Washington, D.C.: Government Printing Office, 1867.

Vanderburg, William O. *Placer Mining in Nevada.* University of Nevada Bulletin vol. 30, no. 4. Geology and Mining Series no. 27. Reno: Nevada State Bureau of Mines and MacKay School of Mines, 1936.

Weed, Walter. *The Copper Handbook: A Manual of the Copper Mining Industry of the World.* Vol. 2. Washington, D.C.: Government Printing Office, 1912.

————. *The Mines Handbook and Copper Handbook.* Vols. 12–17. Washington, D.C.: Government Printing Office, 1916–1926.

North Central Nye County

Angel, Myron. *History of Nevada.* Oakland: Thompson and West, 1881. Reprint, Berkeley: Howell North Books, 1958.

Armstrong, Robert D. *A Preliminary Union Catalogue of Nevada Manuscripts.* Reno: University of Nevada Library and Nevada Library Association, 1967.

Ashbaugh, Don. *Nevada's Turbulent Yesterday: A Study in Ghost Towns.* Los Angeles: Westernlore Press, 1963.

Bailey, Edgar, and David Phoenix. *Quicksilver Deposits in Nevada.* University of Nevada Bulletin no. 41. Reno: Nevada State Bureau of Mines and MacKay School of Mines, 1944.

Ball, Sidney H. *A Geological Reconnaissance in Southwestern Nevada and Eastern California.* U.S. Geological Survey Bulletin no. 308. Washington, D.C.: Government Printing Office, 1907.

Basso, Dave. *Nevada Lost Mines and Hidden Treasures.* Sparks, Nev.: Dave's Printing and Publishing, 1974.

Beebe, Lucius, and Charles Clegg. *U.S. West, the Saga of Wells Fargo.* New York: E. P. Dutton, 1949.

Belmont Courier. 14 February 1874–2 March 1901.

Billeb, Emil W. *Mining Camp Days: Bodie, Aurora, Bridgeport, Hawthorne, Tonopah, Landy, Masonic, Benton, Thorne, Mono Mills, Mammoth, Sodaville, Goldfield.* Berkeley: Howell-North Books, 1968.

Blatchly, A. *Mining and Milling in the Reese River Region of Central and Southeastern Nevada.* New York: Slate & Jones, 1867.

Bowers, Martha, and Hans Muessig. *History of Central Nevada.* Cultural Resource Series no. 4. Tonopah, Nev.: Nye County, 1982.

Browman, Mickey. *Nevada Ghost Town Trails.* N.p.: Abbott & Abbott, 1973.

Brown, Mrs. Hugh. *Lady in Boomtown: Miners and Manners on the Nevada Frontier.* Palo Alto: American West Publishing Company, 1968. Reprint, Reno: University of Nevada Press, 1991.

Bruner, Firmin. *Some Remembers — Some Forgot: Life in Central Nevada Mining Camps.* Carson City, Nev.: Nevada State Park History Association, 1974.

Burt, Olive. *Jedediah Smith: Fur Trapper of the Old West.* New York: Julian Messner, 1951.

Carlson, Helen S. *Nevada Place Names: A Geographical Dictionary.* Reno: University of Nevada Press, 1974.

Central Nevada's Glorious Past. 1978–1996.

Clark, Lewis, and Virginia Clark. *High Mountains & Deep Valleys: The Gold Bonanza Days.* San Luis Obispo, Calif.: Western Trails Publishing, 1978.

Cook, Fred. *Legends of Nye County.* Hawthorne, Nev.: Times, 1978.

Couch, Bertrand F., and Jay A. Carpenter. *Nevada's Metal and Mineral Production, 1859-1940.* University of Nevada Bulletin vol. 37, no. 4. Reno: Nevada State Bureau of Mines and MacKay School of Mines, 1943.

Cronkhite, Daniel. *Ghost Towns of Esmeralda and Nye Counties, Nevada.* Van Nuys: High School Press, 1960.

Davis, Sam P., ed. *The History of Nevada.* Reno: Elms Publishing Company, 1965.

Douglass, William A., and Robert Nylen. *Letters from the Nevada Frontier, Correspondence of Tasker L. Oddie, 1898-1902.* Norman, Oklahoma: University of Oklahoma Press, 1992.

Emmons, William A. *A Reconnaissance of Some Mining Camps in Elko, Lander and Eureka Counties, Nevada.* U.S. Geological Survey Bulletin no. 408. Washington, D.C.: Government Printing Office, 1910.

Ferguson, Henry. *Geology and Ore Deposits of the Manhattan District, Nevada.* Bulletin no. 723. Washington, D.C.: Government Printing Office, 1929.

Florin, Lambert. *Ghost Towns of the West.* New York: Promontory Press, 1971.

Folkes, John. *Nevada's Newspapers: A Bibliography, a Compilation of Nevada History, 1854-1964.* Reno: University of Nevada Press, 1964.

Fox, Theron. *Nevada Treasure Hunters Ghost Town Guide: Handy Reference to Locating Old Mining Camps, Ghost Town Sites, Mountains, Rivers, Lakes, Camel Trails, Abandoned Roads, Springs, and Water Holes.* San Jose: T. Fox, 1961.

Frickstad, Walter N., and Edward W. Thrall. *A Century of Nevada Post Offices, 1852- 1957.* Oakland: Philatelic Research Society, 1958.

Grover, David H. *Diamondfield Jack: A Study in Frontier Justice.* Reno: University of Nevada Press, 1968.

Hall, Shawn. *Romancing Nevada's Past: Ghost Towns and Historic Sites of Eureka, Lander, and White Pine Counties, Nevada.* Reno: University of Nevada Press, 1994.

———. *A Guide to the Ghost Towns and Mining Camps of Nye County, Nevada.* New York: Dodd Mead, 1981.

Harris, Robert P. *Nevada Postal History: 1861 to 1972.* Santa Cruz: Bonanza Press, 1973.

Higgins, L. James. *A Guide to the Manuscript Collections at the Nevada Historical Society.* Reno: Nevada Historical Society, 1975.

Hill, James. *Mining Districts of the Western United States.* Bulletin no. 507. Washington, D.C.: Government Printing Office, 1912.

———. *Mining Districts—California and Nevada.* Bulletin no. 594. Washington, D.C.: Government Printing Office, 1915.

———. *Notes on Some Mining Districts in Eastern Nevada.* U.S. Geological Survey Bulletin no. 648. Washington, D.C.: Government Printing Office, 1916.

Kelly, J. Wells. *First Directory of Nevada Territory (1862).* Los Gatos, Calif.: Talisman Press, 1962.

Kleinhampl, Frank J., and Joseph I. Ziory. *Mineral Resources of Northern Nye County.* Nevada Bureau of Mines Bulletin 99B. Reno: Nevada Bureau of Mines and Geology, University of Nevada, Reno, 1984.

Kral, Victor. *Mineral Resources of Nye County, Nevada.* University of Nevada Bulletin vol. 45, no. 3. Geology and Mining Series no. 50. Reno: Nevada State Bureau of Mines and MacKay School of Mines, 1951.

Labbe, Charles. *Rocky Trails of the Past.* Las Vegas: privately published, 1960.

Lawrence, Edmund. *Antimony Deposits of Nevada.* Nevada Bureau of Mines Bulletin no. 61. Reno: Mackay School of Mines, 1963.

Lewis, Marvin. *Martha and the Doctor.* Reno: University of Nevada Press, 1977.

Lincoln, Francis Church. *Mining Districts and Mineral Resources of Nevada.* Reno: Nevada Newsletter Publishing Company, 1923.

Lingenfelter, Richard E., and Karen Rix Gash. *The Newspapers of Nevada: A History and Bibliography, 1854-1979.* San Francisco: John Howell Books, 1964. Reprint, Reno: University of Nevada Press, 1984.

Mack, Effie. *Nevada: A History of the State from the Earliest Times through the Civil War.* Glendale, Calif.: Arthur H. Clark, 1936.

Manhattan Magnet. 23 March 1917–30 September 1922.

Manhattan Mail. 10 January 1906–24 June 1911.

Manhattan News. 27 January 1906–7 July 1907.

Manhattan Post. 15 October 1910–30 May 1914.

Manhattan Times. 6 July 1907–7 December 1907.

McCracken, Robert. *Bibliography of the Histories of Nye and Clark County.* Tonopah, Nev.: Central Nevada Historical Society, 1991.

Monarch Tribune. 18 August 1906–25 August 1906.

Moody, Eric. *An Index to the Publications of the Nevada Historical Society, 1907–1971.* Reno: Nevada Historical Society, 1977.

Morrissey, Frank. *Turquoise Deposits of Nevada.* Nevada Bureau of Mines Report no. 17. Reno: Nevada Bureau of Mines, 1968.

Morse, Franklin. *The Story of Round Mountain.* Round Mountain, Nev.: Louis Gordon, 1906.

Mountain Champion (Belmont). 3 June 1868–24 April 1869.

Murbarger, Nell. *Ghosts of the Glory Trail.* Palm Desert, Calif.: Desert Magazine Press, 1956.

Nye County Historic Property Survey. Tonopah, Nev.: Nye County, 1980.

Paher, Stanley. *Nevada Ghost Towns and Mining Camps.* Las Vegas: Publications, 1970.

Round Mountain Nugget. 2 June 1906–23 October 1910.

Silver Bend Reporter (Belmont). 30 March 1867–28 July 1868.

Spurr, Josiah. *Geology of Nevada South of the Fortieth Parallel.* Bulletin no. 208. Washington, D.C.: Government Printing Office, 1903.

Stevens, Horace. *The Copper Handbook: A Manual of the Copper Industry of the United States and Foreign Countries.* Vols. 2–10. Washington, D.C.: Government Printing Office, 1908.

Stewart, Virginia W. *Golden Gravel, Manhattan, Nevada in the 1930s.* Morongo Valley, Calif.: Sagebrush Press, 1992.

Stuart, E. E. *Nevada's Mineral Resources.* Carson City, Nev.: State Printing Office, 1909.

Tonopah Miner. 20 June 1902–5 November 1921.

Tonopah Times. 1 December 1915–15 November 1929.

Tonopah Times-Bonanza. 16 November 1929–31 December 1996.

United States Treasury Dept. *Report of J. Ross Browne on the Mineral Resources of the States and Territories West of the Rocky Mountains.* Washington, D.C.: Government Printing Office, 1867.

Vanderburg, William O. *Placer Mining in Nevada.* University of Nevada Bulletin vol. 30, no. 4. Geology and Mining Series no. 27. Reno: Nevada State Bureau of Mines and MacKay School of Mines, 1936.

Weed, Walter. *The Copper Handbook: A Manual of the Copper Mining Industry of the World.* Vol. 2. Washington, D.C.: Government Printing Office, 1912–1913.

———. *The Mines Handbook and Copper Handbook.* Vols. 12–17. Washington, D.C.: Government Printing Office, 1916–1926.

Central Nye County

Angel, Myron. *History of Nevada.* Oakland: Thompson and West, 1881. Reprint, Berkeley: Howell North Books, 1958.

Armstrong, Robert. *A Preliminary Union Catalogue of Nevada Manuscripts.* Reno: University of Nevada Library and Nevada Library Association, 1967.

Ashbaugh, Don. *Nevada's Turbulent Yesterday: A Study in Ghost Towns.* Los Angeles: Westernlore Press, 1963.

Averett, Walter. *Directory of Southern Nevada Place Names.* Las Vegas: privately published, 1962.

Bailey, Edgar, and David Phoenix. *Quicksilver Deposits in Nevada.* University of Nevada Bulletin no. 41. Reno: Nevada State Bureau of Mines and MacKay School of Mines, 1944.

Ball, S. H. *A Geological Reconnaissance in Southwestern Nevada and Eastern California.* Bulletin no. 308. Washington, D.C.: Government Printing Office, 1907.

Basso, Dave. *Nevada Lost Mines and Hidden Treasures.* Sparks: Dave's Printing, 1972.

Beebe, Lucius, and Charles Clegg. *U.S. West: The Saga of Wells Fargo.* New York: E. P. Dutton, 1949.

Belmont Courier. 14 February 1874–2 March 1901.

Billeb, Emil W. *Mining Camp Days.* Berkeley: Howell-North Books, 1968.

Bower, B. M. *The Bellehelen Mine: Bodie, Aurora, Bridgeport, Hawthorne, Tonopah, Landy, Masonic, Benton, Thorne, Mono Mills, Mammoth, Sodaville, Goldfield.* N.p.: Little Brown, 1924.

Bowers, Martha H., and Hans Muessig. *History of Central Nevada.* Cultural Resource Series no. 4. Reno: Nevada State Office, Bureau of Land Management, 1982.

Brooks, Thomas. *By Buckboard to Beatty: The California-Nevada Desert in 1886.* Los Angeles: Dawson's Book Shop, 1970.

Browman, Mickey. *Nevada Ghost Town Trails.* N.p.: Abbott & Abbott, 1973.

Brown, Mrs. Hugh. *Lady in Boomtown: Miners and Manners on the Nevada Frontier.* Palo Alto: American West Publishing Company, 1968. Reprint, Reno: University of Nevada Press, 1991.

Bullfrog District Chamber of Commerce. *Bullfrog Mining District*. Rhyolite: privately published, 1908.

Bullfrog Miner. (Beatty and Bullfrog). 31 March 1905–25 September 1909.

Carlson, Helen S. *Nevada Place Names: A Geographical Dictionary*. Reno: University of Nevada Press, 1974.

Carrara Miner. 1929.

Carrara Obelisk. 8 May 1913–9 September 1916.

Central Nevada's Glorious Past. 1978–1996.

Clark, Lewis, and Virginia Clark. *High Mountains & Deep Valleys: The Gold Bonanza Days*. San Luis Obispo, Calif.: Western Trails Publishing, 1978.

Cook, Fred S. *Legends of Nye County*. Hawthorne, Nev.: Times, 1978.

Cornwall, H. R. *Geology and Mineral Resources of Southern Nye County*. Nevada Bureau of Mines Bulletin no. 77. Reno: Nevada Bureau of Mines, 1972.

Couch, Bertrand F., and Jay A. Carpenter. *Nevada's Metal and Mineral Production*. University of Nevada Bulletin vol. 37, no. 4. Geology and Mining Series no. 38. Reno: Nevada State Bureau of Mines and MacKay School of Mines, 1943.

Cronkhite, Daniel. *Ghost Towns of Esmeralda and Nye Counties, Nevada*. Van Nuys: High School Press, 1960.

Davis, Sam P., ed. *The History of Nevada*. Reno: Elms Publishing Company, 1965.

Elliott, Russell R. *Nevada's Twentieth-Century Mining Boom: Tonopah, Goldfield, Ely*. Reno: University of Nevada Press, 1966.

Emmons, William H. *A Reconnaissance of Some Mining Camps in Elko, Nevada*. U.S. Geological Survey Bulletin no. 408. Washington, D.C.: Government Printing Office, 1910.

Florin, Lambert. *Ghost Towns of the West*. Seattle: Superior Publishing, 1971.

Folkes, John. *Nevada's Newspapers: A Bibliography, a Compilation of Nevada History, 1854–1964*. Reno: University of Nevada Press, 1964.

Fox, Theron. *Nevada Treasure Hunters Ghost Town Guide: Handy Reference to Locating Old Mining Camps, Ghost Town Sites, Mountains, Rivers, Lakes, Camel Trails, Abandoned Roads, Springs, and Water Holes*. San Jose, Calif.: T. Fox, 1961.

Frickstad, Walter N., and Edward W. Thrall. *A Century of Nevada Post Offices, 1852–1957*. Oakland, Calif.: Philatelic Research Society, 1958.

Grover, David. *Diamondfield Jack: A Study in Frontier Justice*. Reno: University of Nevada Press, 1968.

Hall, Shawn. *A Guide to the Ghost Towns and Mining Camps of Nye County, Nevada*. New York: Dodd Mead, 1981.

Harris, Robert. *Nevada Postal History*. Santa Cruz: Bonanza Press, 1973.

Higgins, James. *A Guide to the Manuscript Collections at the Nevada Historical Society*. Reno: Nevada Historical Society, 1975.

Hill, James M. *Mining Districts—California and Nevada*. Bulletin no. 594. Washington, D.C.: Government Printing Office, 1915.

———. *Mining Districts of the Western United States*. Bulletin no. 507. Washington, D.C.: Government Printing Office, 1912.

Kral, Victor. *Mineral Resources of Nye County, Nevada*. University of Nevada Bulletin vol. 45, no. 3. Geology and Mining Series no. 50. Reno: Nevada State Bureau of Mines and MacKay School of Mines, 1951.

Labbe, Charles. *Rocky Trails of the Past*. Las Vegas: privately published, 1960.

Lawrence, Edmund. *Antimony Deposits of Nevada*. Nevada Bureau of Mines Bulletin no. 61. Reno: Mackay School of Mines, 1963.

Lawson, William, and Paul Wesley. *Mines and Mining in Tonopah and Goldfield*. Tonopah, Nev.: privately published, n.d.

Lewis, Marvin. *Martha and the Doctor*. Reno: University of Nevada Press, 1977.

Lincoln, Francis Church. *Mining Districts and Mineral Resources of Nevada*. Reno: Nevada Newsletter Publishing Company, 1923.

Lingenfelter, Richard E., and Karen Rix Gash. *The Newspapers of Nevada: A History and Bibliography, 1854–1979*. San Francisco: John Howell Books, 1964. Reprint, Reno: University of Nevada Press, 1984.

Mack, Effie. *Nevada: A History of the State from the Earliest Times through the Civil War*. Glendale, Calif.: Arthur H. Clark, 1936.

McCracken, Robert D. *Beatty: Frontier Oasis*. Tonopah, Nev.: Nye Country Press, 1992.

———. *Bibliography of the Histories of Nye and Clark County*. Tonopah, Nev.: Central Nevada Historical Society, 1991.

Moffat, James. *Memoirs of an Old-Timer: A Personal Glimpse of Rhyolite, Nevada, 1906–1907*. Tonopah, Nev.: Sagebrush Press, 1963.

Moody, Eric. *An Index to the Publications of the Nevada Historical Society, 1907–1971*. Reno: Nevada Historical Society, 1977.

Morrissey, Frank R. *Turquoise Deposits of Nevada*. Nevada Bureau of Mines Report no. 17. Reno: Nevada Bureau of Mines, 1968.

Murbarger, Nell. *Ghosts of the Glory Trail*. Palm Desert, Calif.: Desert Magazine Press, 1956.

Myrick, David. *Railroads of Nevada and Eastern California*. Vol.2 Berkeley: Howell-North Books, 1963. Reprint, Reno: University of Nevada Press, 1992.

Nye County Historic Property Survey. Tonopah, Nev.: Nye County, 1980.

Paher, Stanley. *Nevada Ghost Towns and Mining Camps*. Las Vegas: Nevada Publications, 1970.

Ransome, Frederick. *Bullfrog District, Nevada*. Bulletin no. 407. Washington, D.C.: Government Printing Office, 1910.

———. *Preliminary Account of Goldfield, Bullfrog, and Other Mining Districts in Southern Nevada*. Bulletin no. 303. Washington, D.C.: Government Printing Office, 1907.

Rhyolite Daily Bulletin. 23 September 1907–8 June 1909.

Rhyolite Herald. 5 May 1905–22 June 1912.

Silver Bow Standard. 5 August 1905–16 January 1906.

Spurr, Josiah. *Geology of Nevada South of the Fortieth Parallel*. Bulletin no. 208. Washington, D.C.: Government Printing Office, 1903.

Stevens, Horace. *The Copper Handbook: A Manual of the Copper Industry of the United States*. Vols. 2–10. Washington, D.C.: Government Printing Office, 1908.

Stuart, E. E. *Nevada's Mineral Resources*. Carson City, Nev.: State Printing Office, 1909.

Tonopah Miner. 20 June 1902–5 November 1921.

Tonopah Times. 1 December 1915–15 November 1929.

Tonopah Times-Bonanza. 16 November 1929–31 December 1996.

Transvaal Miner. 14 April 1906.

United States Treasury Dept. *Report of J. Ross Browne on the Mineral Resources of the States and Territories West of the Rocky Mountains.* Washington, D.C.: Government Printing Office, 1867.

Vanderburg, William O. *Placer Mining in Nevada.* University of Nevada Bulletin vol. 30, no. 4. Geology and Mining Series no. 27. Reno: Nevada State Bureau of Mines and MacKay School of Mines, 1936.

Weed, Walter. *The Mines Handbook and Copper Handbook, 1916–1926.* Vols. 12–17. Washington, D.C.: Government Printing Office, 1916–1926.

———. *The Copper Handbook: A Manual of the Copper Mining Industry of the World.* Vol. 2. Washington D.C.: Government Printing Office, 1912.

Weight, Harold O., and Lucile Weight. *Rhyolite: The Ghost City of Golden Dreams.* Twentynine Palms, Calif.: Calico Press, 1953.

South Nye County

Angel, Myron. *History of Nevada.* Oakland: Thompson and West, 1881. Reprint, Berkeley: Howell-North Books, 1958.

Armstrong, Robert. *A Preliminary Union Catalogue of Nevada Manuscripts.* Reno: University of Nevada Library and Nevada Library Association, 1967.

Ashbaugh, Don. *Nevada's Turbulent Yesterday: A Study in Ghost Towns.* Los Angeles: Westernlore Press, 1963.

Averett, Walter. *Directory of Southern Nevada Place Names.* Las Vegas: privately printed, 1962.

Bailey, Edgar H., and David A. Phoenix. *Quicksilver Deposits in Nevada.* University of Nevada Bulletin vol. 38, no. 5. Reno: Nevada State Bureau of Mines and MacKay School of Mines, 1944.

Ball, Sidney H. *A Geological Reconnaissance in Southwestern Nevada and Eastern Calif.* U.S. Geological Survey Bulletin no. 308. Washington, D.C.: Government Printing Office, 1907.

Basso, Dave. *Nevada Lost Mines and Hidden Treasures.* Sparks, Nev.: Dave's Printing, 1972.

Beebe, Lucius M., and Charles Clegg. *U.S. West: The Saga of Wells Fargo.* New York: E. P. Dutton, 1949.

Belmont Courier. 14 February 1874–2 March 1901.

Billeb, Emil. *Mining Camp Days: Bodie, Aurora, Bridgeport, Hawthorne, Tonopah, Landy, Masonic, Benton, Thorne, Mono Mills, Mammoth, Sodaville, Goldfield.* Berkeley: Howell-North Books, 1968.

Bowers, Martha H., and Hans Muessig. *History of Central Nevada.* Cultural Resource Series no. 4. Reno: Nevada State Office, Bureau of Land Management, 1982.

Brooks, Thomas. *By Buckboard to Beatty: The California-Nevada Desert in 1886.* Los Angeles: Dawson's Book Shop, 1970.

Browman, Mickey. *Nevada Ghost Town Trails.* N.p.: Abbott & Abbott, 1973.

Bullfrog District Chamber of Commerce. *Bullfrog Mining District.* Rhyolite, Nev.: privately published, 1908.

Bullfrog Miner (Beatty and Bullfrog). 31 March 1905–25 September 1909.

Carlson, Helen S. *Nevada Place Names: A Geographical Dictionary.* Reno: University of Nevada Press, 1974.

Carrara Miner. 1929.

Carrara Obelisk. 8 May 1913–9 September 1916.

Central Nevada's Glorious Past. 1978–1996.

Clark, Lewis, and Virginia Clark. *High Mountains & Deep Valleys: The Gold Bonanza Days.* San Luis Obispo, Calif.: Western Trails Publishing, 1978.

Cook, Fred S. *Legends of Nye County.* Hawthorne, Nev.: Times, 1978.

Cornwall, Henry R. *Geology and Mineral Resources of Southern Nye County.* Nevada Bureau of Mines and Geology Bulletin no. 77. Reno: Nevada Bureau of Mines, 1972.

Couch, Bertrand F., and Jay A. Carpenter. *Nevada's Metal and Mineral Production.* University of Nevada Bulletin vol. 37, no. 4. Geology and Mining Series no. 38. Reno: Nevada State Bureau of Mines and MacKay School of Mines, 1943.

Cronkhite, Daniel. *Ghost Towns of Esmeralda and Nye Counties, Nevada.* Van Nuys: High School Press, 1960.

Davis, Sam S. *The History of Nevada.* Reno: Elms Publishing Company, 1965.

Emmons, William H. *A Reconnaissance of Some Mining Camps in Elko, Lander and Eureka Counties, Nevada.* U.S. Geological Survey Bulletin no. 408. Washington, D.C.: Government Printing Office, 1910.

Florin, Lambert. *Ghost Towns of the West.* New York: Promontory Press, 1971.

Folkes, John. *Nevada's Newspapers: A Bibliography, a Compilation of Nevada History, 1854–1964.* Reno: University of Nevada Press, 1964.

Fox, Theron. *Nevada Treasure Hunters Ghost Town Guide: Handy Reference to Locating Old Mining Camps, Ghost Town Sites, Mountains, Rivers, Lakes, Camel Trails, Abandoned Roads, Springs, and Water Holes.* San Jose: T. Fox, 1961.

Frickstad, Walter N., and Edward Thrall. *A Century of Nevada Post Offices, 1852–1957.* Oakland, Calif.: Philatelic Research Society, 1958.

Hall, Shawn. *A Guide to the Ghost Towns and Mining Camps of Nye County, Nevada.* New York: Dodd Mead, 1981.

Harris, Robert P. *Nevada Postal History: 1861 to 1972.* Santa Cruz: Bonanza Press, 1973.

Higgins, L. James. *A Guide to the Manuscript Collections at the Nevada Historical Society.* Reno: Nevada Historical Society, 1975.

Hill, James M. *Mining Districts—California and Nevada.* Bulletin no. 594. Washington, D.C.: Government Printing Office, 1915.

———. *Mining Districts of the Western United States.* Bulletin no. 507. Washington, D.C.: Government Printing Office, 1912.

Kral, Victor. *Mineral Resources of Nye County, Nevada.* University of Nevada Bulletin vol. 45, no. 3. Geology and Mining Series no. 50. Reno: Nevada State Bureau of Mines and MacKay School of Mines, 1951.

Labbe, Charles. *Rocky Trails of the Past.* Las Vegas: privately published, 1960.

Lawrence, Edmund. *Antimony Deposits of Nevada.* Nevada Bureau of Mines Bulletin no. 61. Reno: Mackay School of Mines, 1963.

Lincoln, Francis Church. *Mining Districts and Mineral Resources of Nevada.* Reno: Nevada Newsletter Publishing Company, 1923.

Lingenfelter, Richard E., and Karen Rix Gash. *The Newspapers of Nevada: A History and Bibliography, 1854–1979*. San Francisco: John Howell Books, 1964. Reprint, Reno: University of Nevada Press, 1984.

Mack, Effie. *Nevada: A History of the State from the Earliest Times through the Civil War.* Glendale, Calif.: Arthur H. Clark, 1936.

McCracken, Robert. *Beatty: Frontier Oasis.* Tonopah, Nev.: Nye Country Press, 1992.

————. *Bibliography of the Histories of Nye and Clark County.* Tonopah, Nev.: Central Nevada Historical Society, 1991.

————. *A History of Amargosa Valley, Nevada.* Tonopah, Nev.: Nye County Press, 1990.

————. *A History of Pahrump, Nevada.* Tonopah, Nev.: Nye County Press, 1990.

Moffat, James. *Memoirs of an Old-Timer: A Personal Glimpse of Rhyolite, Nevada, 1906–1907.* Tonopah, Nev.: Sagebrush Press, 1969.

Moody, Eric. *An Index to the Publications of the Nevada Historical Society, 1907–1971.* Reno: Nevada Historical Society, 1977.

Morrissey, Frank R. *Turquoise Deposits of Nevada.* Nevada Bureau of Mines Report no. 17. Reno: Nevada Bureau of Mines, 1968.

Murbarger, Nell. *Ghosts of the Glory Trail.* Palm Desert, Calif.: Desert Magazine Press, 1956.

Myrick, David. *Railroads of Nevada and Eastern California.* Vol. 2 Berkeley: Howell-North Books, 1963. Reprint, Reno: University of Nevada Press, 1992.

Nye County Historic Property Survey. Tonopah, Nev.: Nye County, 1980.

Paher, Stanley. *Nevada Ghost Towns and Mining Camps.* Las Vegas: Nevada Publications, 1970.

Ransome, Frederick. *Bullfrog District, Nevada.* Bulletin no. 407. Washington, D.C.: Government Printing Office, 1910.

————. *Preliminary Account of Goldfield, Bullfrog, and Other Mining Districts in Southern Nevada.* Bulletin no. 303. Washington, D.C.: Government Printing Office, 1907.

Rhyolite Herald. 5 May 1905–22 June 1912.

Spurr, Josiah. *Geology of Nevada South of the Fortieth Parallel.* Bulletin no. 208. Washington, D.C.: Government Printing Office, 1903.

Stevens, Horace. *The Copper Handbook: A Manual of the Copper Industry of the United States.* Vols. 2–10. Washington, D.C.: Government Printing Office, 1908.

Stuart, E. E. *Nevada's Mineral Resources.* Carson City, Nev.: State Printing Office, 1909.

Tonopah Miner. 20 June 1902–5 November 1921.

Tonopah Times. 1 December 1915–15 November 1929.

Tonopah Times-Bonanza. 16 November 1929–31 December 1996.

United States Treasury Dept. *Report of J. Ross Browne on the Mineral Resources of the States and Territories West of the Rocky Mountains.* Washington, D.C.: Government Printing Office, 1867.

Vanderburg, William O. *Placer Mining in Nevada.* University of Nevada Bulletin vol. 30, no. 4. Geology and Mining Series no. 27. Reno: Nevada State Bureau of Mines and MacKay School of Mines, 1936.

Weed, Walter. *The Mines Handbook and Copper Handbook.* Vols. 12–17. Washington, D.C.: Government Printing Office, 1916–1926.

————. *The Copper Handbook: A Manual of the Copper Mining Industry of the World.* Vol. 2. Washington, D.C.: Government Printing Office, 1912.

Weight, Harold O., and Lucile Weight. *Rhyolite: The Ghost City of Golden Dreams.* Twentynine Palms, Calif.: Calico Press, 1953.

Index

Abelman, Nick, 32, 230
Acme, 4
Adams: Jewett, 88, 110, 220; John, 201; Mine, 201
Adaven, 84; Mining and Smelting Co., 37
Adriatic Mine, 221
Advertiser (Ione), 26
Aerolite Mine, 34
Airshaft Mine, 106
Ajax Mine, 190
Ala-Mar Magnesium Co., 92
Albany Mine, 47
Albion Millett Gold Mining Co., 48
Alexander: and Brooklyn Mines Co., 24; Co., 21–23; Mine, 14, 22, 24
Alladin Divide Mining Co., 3
Allis Mine, 12
Allred, 84
Alpha Omega Mine, 12
Amargosa, 234–35, 250, 262. *See also* Original
Amargosa City, 240
Amargosa-Greenwater stage line, 235
American: Borate Co.,

236; Carrara Marble Co., 243; Eagle Mine, 105; Gold Mines Corp., 62; Smelting and Refining Co., 92
Anaconda Mining Company, 35
Ancram, 186, 229
Antarctic Mine, 221
Antelope, 84–85, 102; Mine, 47, 187; Mines Co., 187; Springs, 186–88, 212, 230; View Mine, 187
Apache Hannapah Mines Co., 146
Archer, 2
Arctic Mine, 29
Argent Mining Company, 35
Argentore. *See* Jett
Argonaut Mine, 93
Argus Resources, 160–61, 183
Arista. *See* Carrara
Arista Mining Co., 245
Arizona Mine, 131, 135
Arlington Leaching Works, 34
Arnold, W. D., 221
Arrowhead, 85–87; Annex

Mining Co., 86; Bonanza Mining Co., 87; Consolidated Mining Co., 87; Development Co., 87; Esperanza Mines Co., 87; Extension Mining Co., 86; Inspiration Mines Co., 87; Mine, 86; Mining Co., 86–87; Signal Silver Mining Co., 87; Syndicate Mines Co., 87; Wonder Mines Co., 87; Wonder Mining Co., 87
Ashmeadow Johnnie, 251
Ashmeadows, 235–36
Ash Meadows National Wildlife Refuge, 236
Ashton, 237
Athens, 2–3; Mining District, 2
Atlantic: and Pacific Mining Company, 77; Mine, 219, 221
Atwood, 3, 20; Mine, 3
August Mine, 221
Aurora Bullfrog Mine, 257, 270
Aurum. *See* Original
Austin, 9, 22, 27, 30, 41,

47–48, 131, 134, 148, 154, 166, 221

Baby Mine, 165
Bald Mountain Canyon, 165
Bannock. *See* Hannapah
Barcelona, 41, 126–27, 149, 181, 200
Bare Mountain, 269
Barnes Park. *See* Jackson Mining District
Barnes, Thomas, 29
Barrett, 5; James, 5
Barrick Gold, 267
Bartell: Peter, 166; Silver Mining Co., 166. *See also* Northumberland
Bartlett, George, 48, 221
Basic: Magnesium, 16–17; Ores, 16
Baxter's. *See* Baxter Spring
Baxter Spring, 127–29
Baxter Springs-Manhattan Mining Co., 128
Bay State Mine, 105
Bear Mine, 220
Beard, Clinton, 25
Beatty, 198, 207, 234, 237–39, 243, 260, 263, 268, 272–73; Gold

Mines and Milling Co.,
245; Montillus Murray,
237–39
Beatty Bulletin, 239
Beatty Bullfrog Miner, 238
Beatty Newsbits, 239
Beaty, Alexander, 88–89,
111
Beko, Pete, 109
Bell, T. J., 5, 10–11, 27,
114, 178
Bellehelen, 188–91; Con-
solidated Mines Co.,
190; Development
Corp., 190; Exten-
sion Mining Co., 190;
Merger Mines Co., 189–
90; Mines Co., 190;
Mining Co., 188; Queen
Mine, 190; Ranch, 191
Bellehelen Record, 188
Belmont, 26, 50, 64, 70,
72, 78, 105, 127, 129–
39, 148, 154, 161, 168,
180, 205
Belmont-Austin stage line,
42, 149, 151–52, 163–
64, 165, 166–67, 169,
179, 181
Belmont Big Four Mining
Company, 126
Belmont Courier, 5, 22,
103, 127, 132, 135
Belmont Hook and Ladder
Company, 130
Belmont-Ione stage line,
77
Belmont Mine, 74, 130,
133
Belmont Mining Co., 131,
135
Belmont Security and
Development Co., 136
Belmont Silver Mining
Company, 131
Belmont-Sodaville stage
line, 10, 49, 135
Belmont-Tybo-Eureka
stage line, 103, 108, 110,
161
Belmont-Tybo-Reveille
stage line, 222
Belmont-Westgate stage
line, 81

Ben Hur Mining Co., 63,
142
Berlin, 5–7, 28, 33, 39,
78; Mine, 5, 7, 27; Ich-
thyosaur State Park, 7,
78
Bernice Mine, 142, 203
Best Chance Mine, 62
Betty O'Neal Mine, 20
Big Chief Mine, 9
Big Henry Gold Mining
Company, 12
Bigler Mine, 47
Big Smoky Valley, 40, 140,
154, 158
Big Timber Spring, 271
Birmingham. *See* Jett
Bismark Mine, 114
Black Bear Mine, 226
Black Cabin Well. *See*
Deep Well Station
Black Diamond Mine, 105
Black Dog Mine, 209
Black Hawk Mine, 154
Black Rock Summit. *See*
Silverton
Black Spring, 7–8
Blake's Camp, 191–92
Blanchard, Rev. Benjamin,
164–65
Blizzard Mine, 218
Blue Bell Mine, 166
Bluebird Consolidated
Mining Co., 15
Blue Eagle: Mine, 112,
190; Spring, 88–89
Blue Horse: Mine, 226;
Mining Co., 226
Blue Jacket Mine, 174
Bob. *See* Lodi
Bobby Tungsten Mine, 43
Boggs, Ray, 245
Bonanza, 263. *See also*
Bullfrog
Bonanza Mountain, 266
Bond International Gold,
267
Bonita, 8
Bonnie Clare, 192–95,
218; Bullfrog Mining
Co., 192; Syndicate, 194
Bonnifield, M. S., 61
Booster News, 17
Booth, W. W., 70, 74, 220

Boston Mine, 30, 93
Bowler: Fred, 239; Mine,
239
Bowlerville, 239
Box Springs, 139
Boyd Mine, 29
Bradshaw, Mark, 201
Breyfogle, 139–40; Mine,
139, 251
Bright Star Gold Mining
Co., 150
Bristol, 117
Brohilco Silver Corp., 15,
21
Broken Hills, 56; Silver
Corp., 53, 62
Bromide Hill Mining Co.,
220–21
Brong, Alkali Bill, 234,
259
Brooklyn. *See* Round
Mountain
Brooklyn Mine, 22, 47–48
Brougher, Wilson, 41
Browne's Camp, 240
Brucite. *See* Gabbs
Bruner. *See* Phonolite
Bruner: Henry, 52–53;
Mining District, 51
Bryan Camp. *See* Pioneer
Buccaneer Mine, 101
Buckboard Mine, 69
Buckeye: Mine, 47–48,
102, 148; Mining Co.,
47
Buel: David, 131; Mine,
131
Buffalo: Mine, 141; Mining
Co., 141
Bullfrog, 192, 235, 237,
239–42, 246, 247, 250,
254–56, 260, 262, 265.
See also Original
Bullfrog Gold Bar Mining
Co., 246
Bullfrog-Goldfield Rail-
road, 73, 186, 192–94,
211–12, 218–19, 224,
238–39, 248, 264, 266,
269–70
Bullfrog Gold Reef Mine,
257
Bullfrog Mine, 240, 256

Bullfrog Miner, 238, 240,
263
Bullfrog Mines Co., 257
Bullfrog Mining and Water
Co., 257
Bullfrog Rush Mining Co.,
157
Bullfrog Water Co., 262
Bunker Hill Mine, 209,
229
Bureau of Land Manage-
ment, 94
Burmball Mine, 257
Burns, Tom, 20
Burro Mine, 69
Burt, Chauncey, 37
Butler, 70. *See also* Tono-
pah and Atwood
Butler: Jim, 59–60, 69–
70, 73, 76, 136, 181;
Mine, 4, 46; Ranch, 171
Butterfield: Henry, 89;
Spring, 89–90, 112

Cactus Consolidated
Silver Mines Co., 196
Cactus Leona Silver Corp.,
196
Cactus Nevada Silver
Mine, 195
Cactus Nevada Silver
Mines Co., 196
Cactus Peak, 195–96
Cactus Range Gold Min-
ing Co., 196
Cactus Range Mining Co.,
195
Cactus Silver Mine, 196
Cactus Springs, 186,
195–97
Calico: Mine, 58; Quartz
Mountain Mining Co.,
57
Caliente, 217
California Mine, 102
Camel Springs, 14
Camp Rockefeller. *See*
Cactus Springs
Camp Tungsten, 67. *See
also* Shamrock
Canyon, 243
Carlson Mining and
Milling Co., 209

Carr's Camp. *See* Trappman's Camp
Carr, John "Curley," 230–31
Carrara, 243–46; Marble Co., 126, 246; Portland Cement Co., 245
Carrara Miner, 244
Carrara Obelisk, 243–44
Carrolton, 98, 105
Carson City, 204
Carson, Kit, 64
Casamayou, Andrew, 132
Casket Mine, 114
Cassidy, Butch, 252
Catlin Silver Bow Mining Co., 226
Cedar Mine, 105
Cedar Spring, 197. *See also* Baxter Spring
Centennial: Mine, 30, 32; Mining Co., 30
Central, 140
Central City, 90–91
Centras, John, 8–9
Centrasville, 8–9
Charleston Mine, 38
Charlestown, 246
Chemical and Pigment Co., 206
Chicago Mining and Reduction Company, 41–42
Chief Fraction Mine, 4
Chief Kawich Mine, 215
Chispa Mine, 249, 271
Chloride, 240, 246
Chloride Mining District. *See* Danville
Christmas Gift Mine, 60
Cimmeron. *See* Potomac
Cincinatti Mining Company, 5
Cinnamon Mine, 41
Cirac: Louis, 32; Mine, 61; Mining Co., 61; Ted, 15, 78
Clad Mine, 29
Clair. *See* Bonnie Clare
Clare. *See* Bonnie Clare
Clark Gold Mining Co., 198
Clark: H. H., 240; Jack, 206; Mine, 166; Patsy,

61; Senator William, 264; Tom, 197–98
Clarkdale, 197–98; Consolidated Gold Mines Co., 198; Extension Mining Co., 198; Gold Mines South Extension Co., 198
Clay Camp. *See* Ashmeadows
Clear Creek. *See* Antelope
Clear Creek Mining District, 9
Clifford, 127, 198–200, 211; Edward, 198; James, 198–99; Joe, 142; Thomas, 226
Clifford Extension Gold Mining Co., 198
Clifford Gold Mines Co., 190, 199
Clifford Hill Mining Co., 198
Clifford Mine, 198–200
Clifford Mining Co., 209
Clifford Silver Mines Co., 200
Clifford Silver Mines, Inc., 199
Cliff Spring, 215
Clifton Mine, 112
Clipper Mine, 142
Cloverdale, 9–10, 19, 81
Coen Companies, Inc., 84
Columbia Nevada Mining Co., 188
Columbus Mining District, 80
Columbus-Nevada Mining Co., 189
Columbus-Rexall Mining Co., 80
Combination: Mine, 131, 134; Mining Co., 129, 138
Comet Gold Mining Co., 157
Comstock: Gold Point Mines Co., 92; Merger Co., 150
Congress Mine, 249–50
Consolidated: Golden Arrow Mining Co., 208;

Mayflower Mine Co., 259; Mine, 141
Contact Mining Co., 4
Continental: Mines Co., 94; Silver Mining Co., 220
Cook Ranch, 170
Copon Silver and Gold Mining Co., 151
Copper Range Co., 176
Corcoran: M. W., 179; Ranch, 170
Cornforth Mining Co., 188
Cornworth Mine, 188
Coronation Mining Co., 60
Corrine. *See* Ellsworth
Cotter Mines Co., 208
Courbat Mine, 192
Courter, John, 64
Courthouse Mine, 168
Crabtree, Lotta, 139
Crackerjack Mine, 4
Craig Station, 10
Crane Canyon, 9
Crater Mine, 141
Crescent: Mine, 209, 219, 221; Mining Co., 188
Crockers Ranch. *See* Antelope
Croesus Mine, 257
Cross: Eddie, 260; Ernest, 240, 256
Crow Springs, 70
Crown Point: Globe Mine, 251, 253; Mine, 47; Mining Co., 251
Crown Reef Consolidated Gold Mines Co., 153
Crucible Gold Mining and Milling Co., 202
Crumley: J. G., 226; Newt, 61
Crystal Springs. *See* Pioneer
Cuprite, 218
Currant, 91–92, 103
Currant Creek: Mines Co., 92; Mining District, 91
Cyprus Mines Corp., 168

Daisy Mine, 213
Damon Mine, 61

Daniel Webster Mine, 223
Danville, 84, 93–94; Mining Co., 93
Darrough: Hot Springs, 140–41; James, 140
Davenport. *See* Jett
Davenport, John, 30
Davis, "Diamondfield" Jack, 198, 210
Death Valley, 194, 200, 237; Clay Co., 235; Junction, 200; Railroad, 236; Scotty, 194
Death Valley Magazine, 264
Death Valley Prospector, 264
Deep Well Station, 10–11
DeLamar, 111
DeLamar Lode, 157
Denver Mine, 264
Desert Chief Consolidated Mining Co., 227
Desert Queen Mine, 69, 72, 75, 273
Detroit: Mine, 102; Mining Co., 166
Dexter: Mine, 101; Mining Co., 182; White Caps Mining Co., 182
Diamond Mine, 215
Diamondfield Black Butte Reorganized Mining Co., 44
Diminick Mine, 118
Discovery Mine, 9
Divide, 190; Mining Co., 143
Doctor Mine, 45
Dominion Mine, 101
Donald, Samuel, 21
Dougan, James, 21
Douganville, 21. *See also* Granite
Downey: Mine, 35; Mining Co., 11; Patrick, 18
Downeyville, 10–12, 14; Mine, 11–12; Nevada Mines, 12
Downieville. *See* Downeyville
Duckwater, 95; Indian Reservation, 95

Duluth, 12–13; Gold Mining Co., 12–13
Duluth Tribune, 13
Dutch Mine, 128

Eagle Mining District, 154
Earp, Wyatt, 72
East Belmont, 129–30, 133, 136, 138
Eastern Contact Mine, 272
East Golden. *See* Golden
East Manhattan, 141
Eclipse Mine, 264
Eden, 201–3, 210; Mine, 202
Eden Creek, 90, 210; Mining and Milling Co., 201; Ranch, 202–3
Edgewood. *See* Atwood
Edinburgh Mine, 47
El Dorado: Mine, 131; South Mine, 131; South Mining Co., 135
El Picacho Mining Co., 217
Electrum. *See* Ellendale
Electrum Gold Mines Corp., 205
Elgin-Bellehelen Divide Mining Co., 190
Elizalde Co., 245
Elko, 120
Ellendale, 142, 203–6
Ellendale Star, 204
Ellsworth, 14–15, 78
Ellsworth-Downieville stage line, 10
Elsa Mining Co., 150
Ely, 87, 208
Ely-Goldfield Railroad, 226
Empire Mine, 38
Ernst: George, 170; John, 222; Sara, 102
Esta Buena Mine, 14
Eucalyptus Mine, 93
Eureka, 22, 91, 105, 113–14, 120, 131, 134, 145, 222
Eureka Johnnie Gold Mining Co., 250
Eureka Mine, 105

Eureka-Tybo-Belmont stage line, 97, 109, 139
Everett Mine, 46
Excelsior Twilight Mining Co., 4
Exchequer Quartz Mountain Mining Co., 57

Fairbanks, Dad, 235
Fairbanks Ranch. *See* Ashmeadows
Fairchild: M. D., 221; Oscar, 132
Fairmont Mine, 47
Fairplay. *See* Atwood
Fairplay Mining District, 3
Fairplay Prospector, 4
Fairview Extension Mining Co., 175
Fairview Round Mountain Mines Co., 174
Fairy Mine, 30
Fallini Silver Mining Co., 202
Farcher, Ira, 142
Farrington, Archie, 36, 163
Fisher Mine, 114
Fisherman: Mine, 221–22; Mining Co., 221
Fitch, Thomas, 132
Flagstaff Mine, 15
Florence. *See* Keystone
Florence Mine, 101, 252
Floridin Co., 236
Fluorine, 255–72
FMC Gold Co., 17, 47
Forbes and Nation Mine, 270
Fork's Station. *See* Stonewall
Fountain City. *See* Pioneer
Fourth of July Mine, 30
Fowler Mine, 166
Fraction Mining Co., 60
Francisco, Barney, 19
Franz Hammel Mine, 213
Frazier Springs, 16
Frazier Wells, 16
Fremont, John C., 140
French Republic Mine, 30
Fresno. *See* Georges Canyon

Fresno Mining District, 142
Friends of Rhyolite, 267

Gabbard, John, 230
Gabbs, 16–17, 47, 51, 138
Gabbs, William, 16
Gabbs Exploration Co., 21
Gabbs Gab, 17
Gabbs Independent, 17
Gabbs Valley, 12, 16
Gabbs Valley Enterprise, 17
Gabbs Valley News, 17
Gage, W. S., 33, 66
Garrard. *See* Jett
Garside, Frank, 155–56
Gaston Gold Co., 122
Gates, Humboldt, 155
General Electric Co., 92
General Lee Mine, 14
General Milling Co., 260
Geneva Mine, 209
Geo Drilling Fluids, Inc., 206
George Frawley Ranch, 163
George Martin Mine, 8
Georges Canyon, 142–43; Mining Co., 142
Gettysburg Mine, 47
Giant Mine, 47–48
Giant Silver Mining and Bullion Co., 48
Gibraltar: Gold Mines Co., 245; Mine, 264; Mining Co., 109; North Extension Mine, 32; Silver Hill Mining Co., 31. *See also* Jett
Gila: Consolidated Mining Co., 223; Mine, 221–22; Silver Mining Co., 219–20
Gilbert, 260; Clifford Gold Mines Co., 199; Gilbert International, 187
Giles Camp. *See* Pioneer
Gilleland Mine, 131
Gilroy Mining Co., 245
Gilt Edge. *See* Atwood
Glen Hamilton, 18
Glidden Paint Co., 206
Globe Mining District, 18

Godat, Mary, 116
Golconda Mine, 29
Gold Ace: Annex Mining Co., 245; Consolidated Mining Co., 245; Extension Mining Co., 245; Mine, 245; Mining Co., 245; *See also* Carrara
Gold Bar, 246–47; Mine, 208
Gold Belt. *See* Eden
Gold Canyon, 121, 123
Gold Center, 243, 247–49, 264, 268. *See also* Pueblo
Gold Center News, 248
Gold Coin Mine, 4
Gold Crater, 206–7, 210, 228; Construction and Mining Co., 207; Mining Co., 207
Gold Crown Mining Co., 3, 45
Golden, 19, 44
Golden Anchor Mine, 142
Golden Arrow, 191, 207–10, 221, 225; and Keystone Co., 221; Development Co., 209; Mining Co., 208–9; Mining District, 189; Mohawk Mining Co., 209
Golden Chariot: Mining Co., 213; No. 1 Mine, 213
Golden Eagle Mine, 62–63, 202
Golden Eagle Mining and Milling Co., 13, 53
Golden Fleece Mine, 80
Golden Lion Mining Co., 153
Golden West Mine, 128
Gold Exchange Mine, 45
Goldfield, 64, 72, 121, 142, 187, 192, 198–99, 203, 213, 230, 243, 250, 252–53, 256, 263, 269
Goldfield Auto Stage Co., 241

Goldfield Blue Bell Mining Co., 7
Goldfield Consolidated Mines Co., 186, 217
Goldfield News, 239, 266
Goldfield Quartz Mountain Mining Co., 57, 218
Gold Flat, 210
Gold Frog Mine, 257
Gold Gulch. *See* Transvaal
Gold Hill, 143–44
Gold Hill: Consolidated Mines Co., 143; Development Co., 143; Mining Co., 143
Gold Horn Mine, 145
Gold Metals Consolidated Mines Co., 138
Gold Mountain, 122, 131, 135. *See also* Bonnie Clare
Gold Park, 23, 29; Mining Co., 29
Gold Point, 95
Gold Prince Mining and Leasing Co., 207
Gold Reed: Mine, 215; Mining Co., 215. *See also* Kawich
Gold Reef Mining Company, 20
Gold Ridge Mine, 141
Gold Spring, 121, 123; Mining Co., 122
Gold Wedge Mine, 203
Gold Zone Divide Mining Co., 143
Goldyke, 4, 20
Goldyke Daily Sun, 20
Good Hope Mine, 222
Gooley Mine, 29
Gordon. *See* Round Mountain
Gordon, Louis, 172, 174
Goss Springs, 262
Graham, James, 14, 36, 46
Grand View Mine, 66
Granite, 20–21
Grant City, 90, 96–97
Grant, Ulysses S., 21
Grantsville, 8, 11, 14, 15, 21–25, 33, 39, 126
Grantsville Bonanza, 21

Grantsville Sun, 21
Great Western Mine, 9
Green and Oder Mine, 131
Green Isle Mining District, 148
Green Springs, 25
Gresham Gold Mining Co., 196
Griggs Atwood Mining Co., 3
Grizzly Mine, 41

Halifax-Tonopah Mining Co., 73
Hamilton, 91, 95, 102
Hanks: Carl, 88; Carole, 88
Hannapah, 144–47, 177; Divide Extension Mines Co., 145–46; Extension Mine, 147; Mine, 144; Mining Co., 144; Mining District, 177; Silver Star Mining Co., 146
Happy Kelly Mine, 193
Hard Luck Mine, 192
Hardy, Joe, 201
Harney, Ed, 229
Harriman, 210–11
Harris, Frank "Shorty," 240, 256, 260
Hatterly, J. H., 20
Haven Mine, 12
Havens, P. A., 21, 26, 66
Hawes Canyon, 189, 209, 225–26
Hawthorne, 229
Hawthorne News, 75
Haystack Ranch, 170
Helena. *See* Clifford
Helena Townsite Co., 199
Henry. *See* Bellehelen
Herald, James, 2
Hick's Hot Springs, 211
Hick's Station, 97–98, 110
Hicks, C. W., 97
Highbridge: Consolidated Silver Mining Co., 134; Mine, 131, 133, 135–36
Highland: Boy Mining Co., 221; John, 45
Hiko, 130
Hillside Mine, 148

Hinkle Mine, 257
Hollywood. *See* Carrara
Homestake Mine, 247
Hoodlum Mine, 8
Hooper, Albert, 179
Hornsilver Mining Co., 23
Horseshoe, 211; Mining Co., 188
Horseshutem Springs, 249
Horton, Martin, 88
Hot Creek, 98–100, 109, 179; Consolidated Mines Co., 99, 101; Development Co., 101; Ranch, 104; Ranch Co., 99; Syndicate Trust, 99
Hot Springs. *See* Darrough Hot Springs
Hudson: Leasing Company, 62; Mining and Milling Co., 62. *See also* Royston
Hughes: Howard, 75; Jack, 20
Humphrey: F. G., 132; Frank, 154; John, 154

Ibex Mine, 60
Idlewild, 25; Company, 30; Mine, 30, 149
Ikes Canyon, 179
Illinois: Mine, 35–38; Nevada Mines Corp., 37–38
Imperial Mine, 34
Indian Joe Ranch, 269
Indian: Mine, 38; Mines, Inc., 245; Spring, 212; Springs, 147, 166; Springs Water Co., 261–62
Indiana Mines Exploration Co., 260
Indianapolis Mine, 66
Ione, 14, 21, 26–28, 39, 66–68, 129
Ione-Austin stage line, 8, 18, 80
Ione City. *See* Ione
Ione Gold Mining Co., 27
Ione Valley, 11
Iron Bed Springs. *See* Green Springs

Iron Mercury Mining Co., 68
Irwin. *See* Central City
Irwin, Isaac, 95
Ivy Green Mine, 34

Jack Mine, 12
Jackson: General Stonewall, 227; Mining District, 28–29
Jacksonville, 212
Jacobson Mining and Milling Co., 80
Jamestown, 186–87, 212–14, 228
Jamestown News, 212
Jefferson, 30, 41, 147–51
Jefferson Canyon Consolidated Mining Co., 150
Jefferson Gold and Silver Mining Co., 150
Jefferson Mine, 148
Jefferson Mining Co., 150
Jefferson Silver Mining Co., 149
Jett, 30–32, 109; Canyon, 165, 174, 176; Consolidated Silver Mining Co., 30; John, 30, 165
Jim Graham's Camp. *See* Paradise Peak
JMP Mine, 61
Johnnie, 234–35, 249–51
Johnnie Consolidated Mining Co., 250
Johnnie Mine, 249, 251–53, 271
Johnnie Station. *See* Amargosa
Johnson Mining and Milling Co., 252
Johnson, A. P., 251–52
Johny. *See* Johnnie
Joliet Mine, 222
Jordan and McClellan Mining Co., 157
Jordan: Jack, 229; Michael, 186; Pat, 186
Jumbo Extraction Co., 193
Jumbo Mine, 12, 206
Junction, 151
Juniper Springs, 2. *See also* Athens

Kaiser. *See* Lockes
Kaiser Mine, 105
Kanrohat, Charles, 148–50
Kansas City-Nevada Consolidated Mines Co., 13, 53
Kawich, 170, 210, 214–16, 228
Kawich, Chief, 215
Kawich Consolidated Co., 209
Kawich Gold Mining Co., 215
Kawich Mountains, 214
Kawich Quartzite Mining Co., 215
Kelly's Well. *See* Globe Mining District
Kendall: Tom, 206; Zeb, 121, 229
Kennedy Tellurium Mines Co., 196
Keokuk Mine, 148
Keyser Springs. *See* Lockes
Keystone, 100–2, 121, 209, 221
Keystone Mine, 101, 149
Keystone-Hot Creek Mining Co., 118
King, Charles, 235
King Midas Mining Co., 9
Klondike, 69, 136
Knickerbocker, 32–34; Mining Co., 66; Nevada Mill and Mining Co., 32–33
Kral, Victor, 216
Krotons Mine, 102

Labbe, Charles, 252
Labbe Camp. *See* Johnnie Mine
L.A. Chemical, 236
La Chili Mine, 80
La Plata Mining Company, 47
Ladd, George, 256
Ladd Mountain, 263
Lady Cummings Mine, 166
Lady Washington Mine, 223
Lafayette Mine, 114

Lamont Mine, 12
Landmark Extension Mine, 229
Las Animas Mine, 34
Las Vegas, 268
Las Vegas and Tonopah Railroad, 193–94, 216, 218, 224, 227, 231, 234, 238–40, 243–44, 246–50, 255–56, 259, 264–66, 268, 271
Last Chance Mine, 12, 213
Laughlin, B. V., 269
Laura B. Mine, 60
Lauville, 84, 102
Lawrence Mine, 211
Learnville. *See* Learville
Learville, 151–52, 166
Ledlie, 9
Leeland, 253–54
Lexington Mine, 47
Liberty, 34–35
Liberty Bell Mine, 221
Liberty Group, 34, 34, 54, 64, 221, 222
Liberty-Silver Peak Railroad, 34
Life Preserver Mine, 229
Lillie Mine, 12
Lincoln Gold Mining Co., 195
Lippincott, George, 194
Lisbon Mine, 14
Little Fish Lake Valley, 84, 93, 110
Little Giant Mine, 105
Little Ruth Mining Co., 258
Locke: Elisha, 102; Eugene, 102; Mine, 112
Lockes, 102–3
Lockwood Mine, 33
Lode, The, 204
Lodi, 13, 35–38
Lodi Mines Co., 37
Lodi Tanks. *See* Lodi
Lodivale, 37
Lodi Valley, 11, 53
Logan: Bob, 152, 166; Station, 152
Logan, Thomas, 157
Lognoz Ranch, 151
Lone Jack Mine, 60
Lone Star Mine, 3

Longstreet, 152–54; Jack, 139, 152–53, 235; Mine, 153; Ranch, 208. *See also* Golden Arrow
Lookout Mine, 29
Los Angeles Mine, 35
Louisa Mine, 101
Louisiana Consolidated Mining Company, 118
Lower Town, 98, 103
Lucky Dick Gold Mining Co., 48
Lucky Strike. *See* Ellendale
Lucky Strike Mine, 205–6
Lucky Truck Mine, 197
Lund, 11–12, 16, 20, 37, 111

Ma Alta Mine, 118
MacNamara Mining and Milling Co., 101
Magnolia Mine, 105–6
Magpie Mine, 60
Mammoth. *See* Ellsworth
Mammoth Gold Mining Co., 160
Mammoth Mine, 30, 141
Mammoth Mining District, 10, 14
Manchester Mining Company, 28
Manhattan, 2, 19, 106, 140–41, 154–59, 182–83, 203
Manhattan American Flag Mining Co., 157
Manhattan American Gold Mining Co., 157
Manhattan Big Four Mining Co., 159
Manhattan Calumet Mining Co., 157
Manhattan Catbird Mining Co., 157
Manhattan Century Mining Co., 157
Manhattan Consolidated Mine Development Co., 158
Manhattan Dexter Mining Co., 160
Manhattan Eastern Mining Co., 157

Manhattan Freehold Gold and Silver Mining Co., 154
Manhattan Gold Dredging Co., 160
Manhattan Golden Crust Mining Co., 157
Manhattan Gold King Mining Co., 157
Manhattan Gold Ledge Mining Co., 157
Manhattan Joker Mining Co., 157
Manhattan Magnet, 63, 156
Manhattan Mail, 155
Manhattan Minneapolis Mine, 141
Manhattan Mizpah Mining Co., 157
Manhattan National Mining Co., 157
Manhattan News, 156
Manhattan Placer Co., 160
Manhattan Post, 2, 156
Manhattan Red Top Mining Co., 157
Manhattan Reducing and Refining Co., 158
Manhattan Reliance Mining Co., 157, 160
Manhattan St. Paul, 141
Manhattan Times, 156
Manhattan-Tybo Power Company, 118
Manhattan Union Amalgamated Mines Syndicate, 159
Manhattan United States Mining Co., 157
Manhattan Venture Mining Co., 157
Manhattan-Virginia Mining Co., 157
Manse Ranch, 254
Marble. *See* Lodi
Marble Bluff Mine, 35
Marble Point Mine, 35
Marsh. *See* Georges Canyon
Marsh Mining and Milling Co., 142
Marsh, William, 142, 170

Martin White Company, 113

Martinez, John, 2

Marvel Mining Company, 55

Mary Gray Mine, 29

Marysville, 38–39

Mascot Mine, 209

Maute, Andrew, 21, 132

Mayer, Charles, 230

Mayette, John, 78

Mayflower. *See* Pioneer

Mayflower Mine, 9, 257, 259–60, 270

Mayre Mining Co., 157

Mazie M. Mine, 18

McAllister Mining and Power Syndicate, 150

McCann: Barney, 161; Station, 161–62

McEachin, John, 195

McGonigal Ranch, 170

McKane Mine, 73

McKeehan Station. *See* Craig Station

McMullen, Frank, 226

McNamara, Harry, 3, 99

Meadow Canyon, 148

Mediterranean Mine, 219, 221

Meikeljohn, 255; George, 255

Mellan, 216; Gold Mines, 216; Hazel, 216; Jess, 216

Mercury Mining Co., 28, 67

Metals Reserve Corporation, 35

Mexican Camp, 255

Mexican Mine, 15, 217

Midas, 25, 39. *See also* Ione

Midas Gold Mining and Milling Co., 28

Midland Trail, 106

Midway, 216

Milk Springs Mining District, 114

Millers, 17, 44–45, 71–72, 118, 142, 158, 230

Millett, 48, 162–63; A. B., 2, 46, 48, 162; Christina, 151

Millville. *See* Jefferson

Milton, 39–40

Mineral Hill: Consolidated Mining Co., 141; Mine, 141; Mining Co., 141

Mines Selections Company, 44

Minneapolis Mining Co., 141

Minnimum, Abraham, 151, 163

Minnimum's, 163–64

Miramar Mining Company, 54

Mizpah Hotel, 74

Mizpah Mine, 69, 75

Mobile Mine, 80

Mockingbird Mine, 153

Mogul Mining Company, 61

Mohawk Mine, 154, 213

Monarch, 164–65

Monarch Mine, 29, 47

Monarch Tribune, 164

Monitor. *See* Ellendale or Northumberland

Monitor Mine, 166

Monitor Valley, 69, 152, 167

Monitor-Belmont Mine, 131, 133, 137; Mining Co., 136–37

Montana Mine, 35

Montana Station. *See* Bonnie Clare

Montana-Tonopah Mine, 72

Monte Cristo. *See* Antelope Springs

Monte Cristo Springs, 229

Monterey Mine, 8–9, 105

Montgomery. *See* Johnnie

Montgomery, Bob, 237–39, 261, 264

Montgomery-Shoshone Mine, 264–65

Moore: E. C., 103; H. A., 103, 114; John, 165; Walter, 103; William, 103

Moores Creek Station, 30, 165

Moore's Station, 97, 103–106

Morey, 93, 103, 105–6, 109, 165

Morey District Mining Co., 105

Morey Mine, 106

Morey Mining Co., 105–6

Morgan Creek Ranch, 171

Mormon Charlie, 254

Mormon Well. *See* Nyala

Morning Call Mine, 14

Morning Glory Mining Co., 183

Morristown. *See* Reveille, New

Mount Airy Mine, 105

Mount Charleston, 249, 254

Mount Sterling, 271

Mount Vernon Mine, 14

Mountain Boy Mine, 30

Mountain Champion, 132, 166

Mountain Maid Mine, 128

Mountain Top Mine, 166

Mud Spring, 90; Station, 255

Murphy: John, 40; Mine, 40–41, 43; Thomas, 61

Mustang Mine, 203

Myles, Myrtle Tate, 181

Nancy Hanks Mine, 60

Naneco Resources, 154

National, 44; Bank Mine, 264; Merger Gold Mines Co., 187; Mine, 221

Nature Products Co., 7

Nay, Ellen Clifford, 203

Needles. *See* Arrowhead

Nenzel: Divide Mining Co., 196; Joseph, 196

Nevada Bellehelen Mining Co., 188–89

Nevada Broken Hills Mining Co., 199

Nevada-California Metals Corp., 122

Nevada Central Railroad, 9

Nevada Central Softball League, 168

Nevada Chief Extension Mining Co., 4

Nevada Chief Mine, 4

Nevada Chief Mining Co., 4

Nevada Cinnibar Co., 67

Nevada Co., 5, 11, 18, 27, 29, 33

Nevada Copper News, 4

Nevada Development Co., 225

Nevada Gold Development Co., 175

Nevada Gold Sight Mining Co., 201, 210–11

Nevada Goldfield Mining Co., 192

Nevada Goldfields, Inc., 102

Nevada Manhattan Mining Co., 157

Nevada Massachusetts Mining Co., 16

Nevada Mine, 29

Nevada Mining and Smelting Co., 215

Nevada Operating Co., 257

Nevada Ophir Mining Co., 43

Nevada Ore Reducing Co., 158

Nevada Porphyry Gold Mines Co., 175

Nevada Smelting and Mines Corp., 118

Nevada Southern Railroad, 10

Nevada State Prison, 97

Nevada Wonder Mining Co., 137

Never-Sweat Mine, 189

New Bonnie Clare Mining and Milling Co., 193

New Gibraltar Mines Co., 32

New Gold Bar Mining Co., 247

New Goldfield Sierra Mining Co., 227

New Hope Mining Co., 80

New Life Mine, 61

New Original Bullfrog Mines Co., 257

New Orleans Mine, 80

New Return Mining Co., 15
New York Giants, 150
Newmont Exploration, 54
Newmont Mining Co., 43
Newton-Wolfe Mine, 270
Nic-Silver Battery Co., 194
Nicholl, George, 154
Nickolay Camp. *See* Globe Mining District
Nixon. *See* Gold Flat
Nixon, George, 210, 215, 224
No. 2 Mine, 73
North Divide Extension Mining Co., 62
North Gabbs, 17
North Manhattan, 165–66
North Pole Mine, 4
North Star Mine, 12, 29, 46, 66, 72
North Tonopah King Mining Co., 35
North Twin River Mining District, 47, 162
Northumberland, 147, 152, 166–69; Cave, 169; Mining Co., 168; Ranch, 170
North Union. *See* Jackson Mining District
Nyala, 106–7, 122
Nyala Gold Mines Co., 122
Nye County Land and Livestock Co., 170
Nye County News, 26
Nye Mining Co., 118
Nyopolis. *See* Transvaal

O'Brien's. *See* Wellington
O'Donnell Mountain, 232
Oak Springs, 217, 232; Copper Co., 217
Oasis Bullfrog Mine, 270
Oatman United Gold Mining Co., 4
Oddie, Tasker, 50, 60, 69–70, 170, 198, 210
OK Mine, 128
Okey Davis. *See* Atwood
Okey Davis Mine, 4
Old Colony Mining District, 96

Old Dominion Canyon, 98
Old Dominion Mining Co., 100
Old English Gold Corp., 112
Oldfield, Ed, 251
Olive Mine, 33
Oliver Twist Mine, 101
Olympic Mines Co., 3
Omco, 3
Oneota, 118
Ophir Canyon, 40–43, 77
Ophir Canyon Mining Co., 43
Ophir City. *See* Ophir Canyon
Ore City Mining Co., 271
Original, 256–57
Original Bullfrog Mines Syndicate, 256
Original Horseshoe Mine, 211
Original Mining Co., 160
Original Yellowgold Mining Co., 230
Orion, 262. *See also* Original
Orizaba, 44–45; Mine, 44; Mining and Development Co., 44
Oro Cache Mining and Milling Co., 202
Osterlund Camp, 121
Ostorside. *See* Lockes
Overfield: Mine, 251; Mining Co., 251

Pacific Butte Mines Co., 55
Pacific Coast Borax Co., 200, 253
Pacific States Mining Co., 189
Pacific States Mining Corp., 260
Pactolus, 45–46; Mine, 46; Mining Co., 45
Page: Claudet, 207; Marl, 207
Pahranagat stage line, 130
Pahrump Valley, 249
Pahute Mesa, 206–7, 210, 212, 229

Palmer's Well. *See* Rose's Well
Palo Alto, 154
Paradise Peak, 46–47; Mining Co., 46
Park Canyon, 47–49, 55, 149–50, 162
Parker: George Cassidy, 252; Robert Leroy, 252
Paymaster Mine, 52–53
Peavine, 49–51; Creek, 160; Johnny, 198
Penelas, 51; Mine, 51; Mining Co., 51; Severino, 51
Peoria Mine, 14
Periscope Mine, 229
Perkup. *See* Ellendale
Petersgold, 218
Peterson Mine, 29
Phoenix Mine, 33–34
Phonolite, 12–13, 51–54
Phonolite Paymaster Mining Co., 53
Phonolite Silent Friend Mining Co., 53
Phonolite Townsite, Water & Light Co., 52
Pine Creek, 169–71
Pine Nut Mine, 141
Pioche, 106, 111, 114
Pioneer, 239, 257–60, 269
Pioneer Consolidated Mines Co., 259–60
Pioneer Mine, 257, 259–60, 270
Pioneer Press, 258
Pioneer Topics, 258
Piute Mine, 35
Plymouth Goldfield Mining Co., 195
Point of Rocks Mine, 106
Polaris Mine, 264
Polygamy Well. *See* Nyala
Potomac, 34, 54–55; Mine, 34, 64
Potts, 171–72; William, 171
Pritchard's Station, 97, 108
Prussian Mine, 148–49
Prussian Troy Mine, 148
Pueblo, 55
Pueblo Gold Mines Co., 55

Punch Bowl Ranch, 170
Puritan and Gold Reef Mining Co., 258
Putney Gold Mining Co., 166
Pyrennes Mine, 80

Quartz Mountain, 56–59, 218
Quartz Mountain Metals Co., 57
Quartz Mountain Metals Mine, 58–59
Quartz Mountain Miner, 56
Quartz Mountain Mines and Milling Co., 57
Quartz Mountain Townsite Co., 56
Queen City. *See* Mellan
Queen Liliuokalani, 48
Queenie Consolidated Mining Co., 150
Quigley Reduction Co., 193
Quincy. *See* Royston
Quinn Canyon Mountains, 121
Quintero Co., 151–52, 166
Quintero Mine, 133

Railroad Valley, 88–89, 91, 106–7, 223
Railroad Valley Co., 90
Ralston, 187, 218–19
Rambler-Tonopah Mining Co., 180
Ramsey, Harry, 60
Rattler: Mine, 18; Mining Co., 18
Rattlesnake. *See* Keystone
Rattlesnake: Canyon, 98; Mine, 192
Rawhide, 56
Ray, 16, 59–61
Ray and O'Brien Gold Mining Co., 59–60
Ray Consolidated Gold Mining Co., 61
Ray Extension Mining Co., 60
Ray, Judge L. O., 12, 16, 59–60
Ray Mining Co., 60

Ray-Tonopah Mining Co., 61
Raymond and Ely Mining Co., 35
Raymond Van Ness Mining Co., 181, 205
Red Hill Florence Mining Co., 138
Red Indian Metallic Paint Co., 228
Red Lion Consolidated Mines Co., 227
Reed, O. K., 170, 191, 214–15, 232
Reed Mining District, 191
Reese River Gold Ledge Mining Co., 8
Reese River Gold Mining Co., 8
Reese River Reveille, 41, 105, 221
Reilly, Frank, 186
Reliance Mine, 61
Reno, 148
Reno Evening Gazette, 271
Reno Goldfield Mining Co., 168
Reorganized Original Bullfrog Mines Syndicate, 256
Reorganized Pioneer Consolidated Mines Co., 260
Republic, 45, 61–62
Return: Mine, 15; Mining Co., 15
Reveille, 23, 87, 102, 118, 121, 209, 217, 219
Reveille (Gila) Mill, 219–20
Reveille-Liberty Mining Co., 221
Reveille Mining and Milling Co., 221
Reveille, New, 220–21, 223
Reveille, Old, 219–23
Reveille-Tonopah Mining Co., 220
Reveille Valley, 221
Revenue Mine, 80
Revive. *See* Northumberland
Rex Mine, 60
Rhyolite, 193, 204, 234, 237, 239–43, 246, 248–49, 254–57, 259–68, 272
Rhyolite Consolidated Mines Co., 266
Rhyolite Daily Bulletin, 263
Rhyolite Herald, 238, 242, 263, 266
Rich Hills Mining District, 46
Richmond Mine, 93
Rieschke, Herman, 221
Rigby Mine, 34
Right Tip Mine, 9
Rochester, 196
Rocket: Mine, 46; Mining Co., 195
Rosalie Mining Company, 61
Rosario Mining and Milling Co., 158
Rose's Well, 268
Rosecrans, General, 14
Roswell. *See* Rose's Well
Round Mountain, 19, 48, 143, 172–77, 247
Round Mountain Blue Jacket Mining Co., 175
Round Mountain Gold Corp., 176
Round Mountain Gold Dredging Co., 176
Round Mountain Mines Co., 174–75
Round Mountain Mining Co., 48, 174–75
Round Mountain Nugget, 172
Round Mountain Sphinx Mining Co., 174
Royston, 62–63
Royston Blue Turquoise Mines Co., 63
Royston Coalition Mines Co., 63
Royston Piedmont Mining Co., 63
Royston Turquoise Mines Co., 63
Royston, W. H., 62
Rural Mines, Inc., 37
Russell, George, 104
Rustler Mine, 122
Rye Patch, 177

Sage Hen Mine, 93
Sailor Boy Mine, 148
San Antone. *See* San Antonio
San Antonia. *See* San Antonio
San Antonio, 64–65, 147
San Antonio Mining Co., 34
San Antonio Silver Mining Co., 30
San Carlos, 224
San Felipe Mining Co., 57
San Francisco Mine, 29, 102
San Juan, 65–66
San Juan Canyon, 65
San Lorenzo. *See* Potomac
San Lorenzo Mine, 54
San Pedro, Manuel, 14, 21, 23, 126
San Rafael Consolidated Mines Co., 57
San Rafael Development Co., 57
San Rafael Mine, 58
Sand Mound Mine, 35
Sandstorm Mine, 12, 203
Santa Fe Mine, 221
Santa Fe Railroad, 249
Santa Rosa Mine, 47
Santo Nino Mine, 80
Sarah Mine, 12
Sarcobatus Flat, 212
Saulsbury Wash, 203
Savage Mine, 12
Sawmill, 109
Scheebar: Mercury Mine, 46; Syndicate, 46
Scheel, Charles, 162
Schultz, Fred, 213
Schwab, Charles, 198, 220, 237, 246, 264–65
Sciota Mine, 189
Sciota Mining Co., 188
Sciota-Nevada Mining Co., 189
Scotty's Castle, 194–95, 200
Scotty's Junction. *See* Death Valley Junction
Searville. *See* Learville
Seay, George, 90

Sedan Mine, 101
Seventy-six Mine, 30
Sewell, Harvey, 171
Seyler Ranch, 154
Seymour. *See* Central City
Shamberger, Hugh, 230
Shamrock, 66–68; Canyon, 28; Mines Co., 67
Sharp, 84. *See also* Adaven
Sharp: George, 107; Lewis, 88; Lina, 88; M. S., 103; Mary, 107; Thomas, 84
Sharp Ranch. *See* Nyala
Sheba Mine, 179
Shermantown, 132
Shoshone. *See* Round Mountain
Shoshone Indian Reservation, 95
Shoshone Mine, 38
Shoshone Quicksilver Mining Co., 67
Sierra Magnesite Co., 17
Sierra Nevada Mine, 148
Sierra Pacific Power Co., 161
Sierra Vista Mines Co., 46
Silver Arrowhead Mining Co., 87
Silver Bar Mine, 165
Silver Bend Mining Co., 133
Silver Bend Reporter, 132
Silver Bow, 191, 224–27
Silver Bow Belle Mining Co., 226
Silver Bow Consolidated Mining Co., 226
Silver Bow Mining and Milling Co., 226
Silver Bow Standard, 224
Silver Buck Mine, 221
Silver City, 3
Silver Divide Mines Co., 55
Silver Dollar Mine, 190
Silver Dyke Mine, 106
Silver Extension Mining Co., 144
Silverfield Mine, 201
Silverfields Ajax Mines Co., 190

Silver Glance, 177–78; Mine, 177; Mining Co., 178. *See also* Hannapah
Silver Hills Extension Mining Company, 109
Silver Hoard Mining Co., 226
Silver Horn Mine, 111
Silver Leaf Mine, 21, 145, 147
Silver Mines Selection Company, 45
Silver Mount Mine, 166
Silver Palace Mine, 24
Silver Point. *See* Jett
Silver Point Mine, 31, 147
Silver Star Mine, 9
Silver Top Mine, 69, 75
Silver Wave Mine, 14
Silverton, 31, 109; Mine, 109
Silverzone. *See* Hannapah
Silverzone Extension Mining Co., 177
Silverzone Mines Co., 145, 177
Silverzone Mohawk Co., 145
Slavin, Ed, 232
Smith, Borax, 238
Smith, Guy, 16
Smith's Station. *See* Stone House
Smoky Valley, 151–52, 167, 181
Smoky Valley Gold Mining and Milling Co., 48
Smoky Valley Mining Co., 176
Smuggler Mine, 106
Snow King Mine, 9
Sodaville, 72, 135–36
South Bullfrog, 269
South Gabbs, 17
South Pole Mine, 4
South Prussian Mine, 148
Southern Pacific Railroad, 72
Southey, A. B., 218
Southgold Mine, 202
Southgold Nevada Mines Co., 202
Southwestern Nevada Co., 220

Spanish Belt: Consolidated Silver Mines Co., 126; Extension Mining Co., 126; Mining District, 126
Spanish Spring, 178; Mining Co., 178
Sparrow Mine, 12
Springdale, 259, 269–70
Springdale Mining and Milling Co., 270
Springer, Harry, 16, 206
Springfield: Mine, 34; Mining District, 179
St. Charles Mine, 149
St. Elmo. *See* Sumo
St. Helena Mine, 80
St. Louis: Mine, 47; Mining Co., 93
St. Paul: Mine, 47, 128; Mining Co., 141
Star of the West: Mine, 29; Mining Co., 29
Stargo, 84, 102, 110
Starlight Mine, 257
Stephanite. *See* Silver Bow
Stephenson, F. S., 16
Sterlag Tunnel, 228
Sterling. *See* Stirling
Sterling: Mine, 271; Mining Co., 271
Stewart, William, 241–42, 263
Stimler, Harry, 31–32, 70, 109, 142, 161, 170, 218, 245
Stirling, 271
Stokes, J. Phelps, 5, 11, 18, 27, 29, 33
Stone Cabin Ranch, 142, 209
Stone House, 179–80
Stoneham, Charles, 150
Stonewall, 227–28; Mine, 227; Mountain, 207, 227
Storm Cloud Mine, 191–92
Storm King Mine, 27
Stott, J. R., 2
Stranger Mine, 80
Stratford, 12. *See also* Duluth
Sullivan Mine, 18

Sulphide, 228–29
Sultan Mine, 47
Summa Corp., 75
Summerville. *See* Bonnie Clare
Summit Canyon, 47
Summit City. *See* Ellsworth
Summit Mine, 29, 47
Summit Station, 110
Sumo, 180
Sun Battery Co., 194
Sunnyside, 110–11
Sunnyside Mine, 149, 174
Sunrise Mine, 18, 91
Sunset Mining and Development Co., 265
Super Six Mining Co., 63
Swastika. *See* McCann Station
Sylvanite. *See* Hannapah

Tampico Mine, 38
Tate, Thomas, 42, 181
Tate's Station, 181
Taylor. *See* Blake's Camp
Tecumseh Mine, 47
Telluride, 255, 272
Tenneco, 236
Tent City, 17
Terrell, Clyde, 238–39
Texota Oil Co., 89
Thatcher, Roy, 145
The Willows. *See* Sawmill
Thompson Mine, 197
Thorp. *See* Bonnie Clare
Thorp's Wells. *See* Bonnie Clare
Tiger Mine, 142, 203
Timber Mountain. *See* Stirling
Timber Mountain Mining District, 271
Tognoni, Joe, 109
Toiyabe. *See* Gabbs
Toiyabe City. *See* Ophir Canyon
Toiyabe Mining Co., 9
Toiyabe Mountains, 9, 162–63
Toiyabe Treasure, 17
Toledo-Tonopah Mining Co., 61
Tolicha, 186, 229–30

Tom Burns Camp. *See* Goldyke
Tomahawk Mining Co., 188
Tonogold, 68–69
Tonogold Townsite Co., 68
Tonopah, 59, 64, 68–77, 86, 118–19, 128, 136, 145–46, 158, 174, 177, 198, 200, 203–4, 208–9, 214, 220, 224–25, 229, 263
Tonopah and Goldfield Railroad, 71–75, 187, 264
Tonopah and Tidewater Railroad, 229, 234–35, 236–37, 238–39, 248, 253, 264, 273
Tonopah Banking Corp., 170
Tonopah-Belmont Mining Co., 71, 143
Tonopah Bonanza, 70, 74, 220
Tonopah-Brohilco Mines Corp., 15
Tonopah-Clifford-Reveille stage line, 199, 220
Tonopah Daily Bonanza, 74–75
Tonopah Daily Times, 75
Tonopah Extension: Mine, 72; Mining Co., 73, 142
Tonopah Gold and Copper Mining Co., 180
Tonopah Gold Zone Mining Co., 143
Tonopah Ione Mining, Milling, and Leasing Co., 28
Tonopah-Jupiter Mining Co., 68
Tonopah-Kawich Mining Co., 189
Tonopah-Las Vegas stage line, 198
Tonopah Liberty Mining Co., 34
Tonopah-Manhattan stage line, 178
Tonopah Mines Syndicate, 45

Tonopah Mining Co., 54, 70–71, 73, 143
Tonopah Northern Mining Co., 180
Tonopah Republic Mining Co., 62
Tonopah Sun, 74
Tonopah Times-Bonanza, 75, 239
Tonopah Wonder Mining Co., 34
Top Notch Mine, 221
Toquima Mountains, 129
Toyah, 77
Tramp Mine, 264–65
Transvaal, 272–73; Mine, 272
Transvaal Miner, 273
Transvaal-Nevada Mining Co., 272
Transvaal Tribune, 273
Transylvania Mine, 131
Trappman, Hermann, 230
Trappman Mining Co., 230
Trappmans, 187
Trappman's Camp, 228, 230–31
Treadwell-Yukon Mining Co., 102, 118–19
Treasure Hill: Mine, 109; Mining Co., 109
Triangle Mines Co., 62
Tripod Mine, 145
Trippel, Alexander, 93
Troy, 89–90, 111–13
Troy Mine, 47, 112
Troy Silver Mining Co., 113
Twilight Mine, 34
Twin River. See Ophir Canyon
Twin River Giant Mine, 47
Twin River Mining Co., 40–41
Twin River Ranch, 162
Twin Springs, 114
Twin Springs Ranch, 103
Two-G Mine, 114–15
Tybo, 8, 23, 90–91, 98, 102–3, 105, 109, 114–20, 161, 215
Tybo Consolidated Mining Co., 98, 114–18

Tybo Dominion Mines Co., 102, 118
Tybo Mine, 118
Tybo Sun, 115
Tyke Mine, 158

Uncle Sam Mine, 190
Union, 23, 33, 77–79, 135
Union Amalgamated Mining Co., 158
Union Canyon, 6, 15
Union Mine, 35, 149
United Cattle and Packing Co., 170
United Lodi Mines Co., 37
United States Machinery Co., 119
United States Milling Co., 239
Upper Weston. See Ellsworth
Ural Mining Co., 33
Uranium and Federated Minerals Co., 147

Valley View Mine, 69, 257
Valparaiso Mine, 80
Van Ness, 181–82
Vanderbilt Minerals Co., 184
Veatch, George, 33, 78
Victor Mine, 73
Victorine Mine, 221
Virgin Gold Mining Co., 226
Volcano. See Hannapah
Vulcan Mine, 34

Wadsworth, 11, 14, 22, 35, 148
Wagner, 224, 231
Walk, Bob, 223
Warburton, Thomas, 30
War Eagle: Mine, 29; Mining Co., 182
War Path Mine, 60
Ward, 113
Warm Springs, 103, 120–21
Warm Springs–Eureka stage line, 97
Warner: Mine, 79–80; Mining and Milling Co., 80

Warrior: Consolidated Gold, 3; Gold Mining Co., 2; Mine, 2–3
Washington, 65, 79–81; Canyon, 65; Mining Co., 80
Webster: Mine, 24; Mines Corp., 24
Weepah, 168
Wellington, 10, 228, 231; Mine, 257
Wells Fargo, 81, 109, 114, 127, 148–49, 155, 253
West End Consolidated Mining Co., 73
West Spanish Belt Silver Mining Co., 126
West Toledo Mining Co., 80
West Uranium Mines Corp., 138
Western Leadfield Co., 63
Western Magnesium and Chemical Corp., 16
Western States Mineral Corp., 169
Western Union, 134, 258
Westgate, 127
Westgate-Ione-Belmont stage line, 64
Westvaco Chlorine Products Corp., 92
Wheelbarrow Peak, 212
Whipple, John, 111
Whipple Ranch. See Sunnyside
White Caps, 182–84
White Caps Extension Mines Co., 183
White Caps Gold Mining Co., 183
White Caps Mine, 158, 161, 183–84
White Caps Mining Co., 158, 182–83
White Pine Mining Co., 37
White Pine Telegram, 132
White River, 91
White River Valley, 110
White Rock Spring, 232
White's Ranch. See Manse Ranch
Wide West Mine, 47

Williams, Joseph, 99
Willow Creek, 121–23; Gold Mining and Milling Co., 121–22; Mining Co., 121–22
Willow Creek Ranch. See Sawmill
Willow Springs, 81
Willow Springs Mining Co., 180
Wilson Divide Mining Co., 62
Wilson Ranch, 171
Wilson's Camp, 228, 232
Wilson, Lloyd, 206
Wingfield, George, 12, 31, 37, 57, 121–22, 128, 161, 186–88, 191, 198, 205, 210, 224, 229, 259, 272
WJP Mine, 61
Wonder Mine, 87
Woodbridge Mine, 149
Workman, Ed, 44
World Exploration Co., 146
Wyoming-Scorpion Mine, 231

Yankee Blade, 47
Yeiser, Ed "Jumbo," 229–30
Yellow Jacket Mining Co., 245
Yellow Tiger Consolidated Mining Co., 227
Yellow Tiger Mining Co., 227
Yellowgold. See Trappman's Camp
Yellowgold Consolidated Mining Co., 231
Yellowgold Mine, 230
Yellowgold Mining and Milling Syndicate, 230
Young Camp. See Johnnie Mine
Young, Joseph, 254
Yount's Ranch. See Manse Ranch

Zabriskie, John, 139
Zanzibar Mining Co., 160
Ziegler Mine, 257